# Skeletal Muscle from Molecules to Movement

## A Textbook of Muscle Physiology for Sport, Exercise, Physiotherapy and Medicine

**David Jones**
School of Sport and Exercise Sciences
University of Birmingham
Edgbaston, Birmingham, UK

**Joan Round**
School of Sport and Exercise Sciences
University of Birmingham
Edgbaston, Birmingham, UK

**Arnold de Haan**
Institute for Fundamental and Clinical Human Movement Sciences
Faculty of Human Movement Sciences
Vrije Universiteit
Amsterdam, The Netherlands

CHURCHILL
LIVINGSTONE

EDINBURGH LONDON NEW YORK OXFORD PHILADELPHIA ST LOUIS SYDNEY TORONTO 2004

CHURCHILL LIVINGSTONE
An imprint of Elsevier Limited

**First published 2004**
ISBN 0 443 07427 5

**British Library Cataloguing in Publication Data**
A catalogue record for this book is available from the British Library

**Library of Congress Cataloging in Publication Data**
A catalog record for this book is available from the Library of Congress

**Notice**
Medical knowledge is constantly changing. Standard safety precautions must be followed, but as new research and clinical experience broaden our knowledge, changes in treatment and drug therapy may become necessary or appropriate. Readers are advised to check the most current product information provided by the manufacturer of each drug to be administered to verify the recommended dose, the method and duration of administration, and contraindications. It is the responsibility of the practitioner, relying on experience and knowledge of the patient, to determine dosages and the best treatment for each individual patient. Neither the Publisher nor the editors assume any liability for any injury and/or damage to persons or property arising from this publication.

**The Publisher**

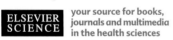

ELSEVIER
SCIENCE

your source for books, journals and multimedia in the health sciences
**www.elsevierhealth.com**

The publisher's policy is to use **paper manufactured from sustainable forests**

Printed in China by Elsevier

# Contents

# Preface

Muscle is an important, interesting and popular subject for students from a range of disciplines including movement sciences, sport science, physiotherapy, biochemistry and medicine, and it is our experience of teaching in all of these areas that has led us to produce this book. It is intended to provide an accessible source of basic information as well as an introduction to a number of areas of research interest. Chapters 1–6 and 15 provide the groundwork, covering structure and function, biochemistry and the mechanism of force generation and we hope that this will be useful both for students coming new to the subject or for those who wish to revise and be brought up to date. The other chapters deal with topics where there is controversy and speculation and we have tried to summarize the current state of knowledge, to point out what is well established, what is speculative and where current research is leading. These latter sections reflect our own interests and research and, hopefully, also our enthusiasm for the subject.

The learning process is complex and requires more than a passive downloading of facts. Consequently we have added a series of questions at the end of each chapter to stimulate a more active learning process. Some of the questions are factual, based on the material in that chapter, others refer back to previous chapters and others refer forward. For a few questions there is no correct answer, only speculation, and we hope these will illustrate the limitations of our knowledge and the need for further research.

The figures and illustrations are reproduce here in black and white or as halftones with the intention of keeping the cost of the book to a minimum and making it affordable for all. However, colour certainly enhances illustrations, especially histology sections, and we have therefore included a CD with all the illustrations in their original colours and hope that this will complement the monochrome text.

It is said that you never understand a subject until you come to teach it and this is even truer when it comes to writing a book such as this where we have continued to debate and refine the text even to the final proof stage. In writing this book we have learned a lot and hope that we can pass this knowledge on to new generations of students.

*David Jones*
*Joan Round*
*Arnold de Haan*
*August 2003*

# Acknowledgements

A great deal of the material in this book is based on research carried out with close colleagues and the references in the text indicate the importance to us of these collaborations. In particular we would like to thank Richard Edwards and Tony Sargeant for their support, encouragement and lively debate over the years. One of us would also like to remember David Hill who was a major influence at a time when he (DAJ) was just starting to take an interest in muscle physiology.

# Abbreviations

| | | | |
|---|---|---|---|
| ACh | acetylcholine | MEPP | miniature end-plate potential |
| AChR | acetylcholine receptor | MHC | myosin heavy chain |
| A-CSA | anatomical cross-sectional area | MLC | myosin light chain |
| ADP | adenosine diphosphate | mRNA | messenger ribonucleic acid |
| AMP | adenosine monophosphate | MRS | magnetic resonance spectroscopy |
| ATP | adenosine triphosphate | MVC | maximum voluntary contraction |
| BCAA | branched-chain amino acid | NAD(H) | (reduced) nicotinamide–adenine |
| CGRP | calcitonin gene-related peptide | | dinucleotide |
| CK | creatine kinase | NADH-TR | NADH tetrazolium reductase |
| CoA | coenzyme A | NMJ | neuromuscular junction |
| CPT | carnitine palmitoyl transferase | P | phosphate |
| DHP | dihydropyridine | PCr | phosphocreatine |
| DMD | Duchenne muscular dystrophy | P-CSA | physiological cross-sectional area |
| DOMS | delayed-onset muscle soreness | PDH | pyruvate dehydrogenase |
| EC | excitation–contraction (coupling) | | (complex) |
| EDTA | ethylenediamine tetra-acetic acid | PFK | phosphofructokinase |
| EM | electron miscroscopy | $P_i$ | inorganic phosphate |
| EMG | electromyography | PTP | post-tetanic potentiation |
| $FAD(H_2)$ | (reduced) flavin–adenine | SCI | spinal cord injury/injured |
| | dinucleotide | SDH | succinate dehydrogenase |
| GABA | γ-aminobutyric acid | SMA | spinal muscular atrophy |
| GTP | guanosine triphosphate | SR | sarcoplasmic reticulum |
| HIF | hypoxia-inducible factor | SREC | short-range elastic component |
| IMP | inosine monophosphate | SSRI | selective serotonin-reuptake |
| $IP_3$ | inositol 1,4,5-triphosphate | | inhibitor |
| IU | international unit | TCA | tricarboxylic acid |
| ME | myalgic encephalopathy | Tm | tropomyosin |

# 1

# Structure of the muscle fibre

One of the major differences between plants and animals is that, whereas green plants can obtain energy from sunlight and nutrients from the soil they grow in, animals have to forage for food or prey. Purposeful movement is, therefore, characteristic of higher animals, and this function requires the activity of voluntary skeletal muscle. The consequences of skeletal muscle activity are very evident—be it in the gentle movements of quiet breathing or the more dramatic endeavours of an athlete sprinting or jumping. Muscle is one of the few tissues in which it is possible to understand, at the molecular level, how these very different actions take place and are regulated. The main purposes of this and following chapters is to explain the link between molecular mechanisms and the function of the whole muscle working in the body.

One of the reasons why so much is known about the molecular mechanisms underlying function in skeletal muscle is because of the highly ordered structure of the contractile elements. The description that follows begins with the contractile proteins, explains how these are arranged into sarcomeres and myofibrils within muscle fibres, and how the fibres then combine with connective tissue to form the whole muscle.

## THE CONTRACTILE PROTEINS

Although deliberate movement is one of the features that separates animal from plant life, nevertheless contractile proteins, both myosin and actin, are found in all types of cell, being responsible for protoplasmic streaming and movement of intracellular organelles. Actin is a protein of great antiquity and is highly conserved in the sense that actins from animal and plant cells are functionally and immunologically similar. Skeletal and cardiac muscle are unusual, not for possessing actin and myosin, but for their particularly high content (about 80% of total protein), and for having these two proteins arranged in a highly ordered array within the cell, permitting the controlled generation of force and movement.

## Actin

Actin is a globular protein (G-actin) with a molecular weight of 42 000, which polymerizes into double helical strands (F-actin) (Fig. 1.1). The polymerization of actin involves splitting ATP and binding of ADP, which constitutes about 90% of the total ADP in muscle. The actin filaments (also known as *thin filaments*) are variable in length, with mammalian filaments being somewhat longer than those of amphibian muscle. The thin filament length also varies between muscles in the same animal and even within a sarcomere so that the edge of the I band (see Fig. 1.4) may be somewhat irregular. Tropomyosin and the three troponin molecules, TnC, TnT and TnI, form the other constituents of the thin filaments (Fig. 1.1). The tropomyosin extends over seven actin subunits, blocking the sites where myosin can bind to the thin filament until caused to move by calcium binding to troponin C.

The actin filaments join at one end to form the Z-line structure. At the Z line, the actin filaments are in a square array, with each thin filament in one half-sarcomere being linked to four other filaments in the next half-sarcomere, with the protein α-actinin forming the connections between the actin filaments. In fast muscles, the linkage is quite simple, giving a thin Z line, whereas in slow fibres there may be several connections between the two sets of thin filaments, giving a thicker Z line (see Figs 4.12 & 11.1).

## Myosin

As with actin, myosin is found in animal and plant kingdoms, the variety found in mammalian skeletal muscle being known as myosin II. The molecule consists of two identical chains, each with a molecular weight of approximately 200 000, together with four light chains of around 20 000 molecular weight each (for reviews see Rayment et al 1993a,b). In mollusc muscle the myosin light chains have a clear regulatory role, binding calcium and controlling the activity of myosin. Mammalian light chains can substitute for mollusc light chains, demonstrating that they have functional potential but, to date, no unequivocal role for these proteins has been found. The composition of the light chains differs between fast and slow muscles.

The myosin molecule can be split into two major fragments (Fig. 1.2A). The globular head,

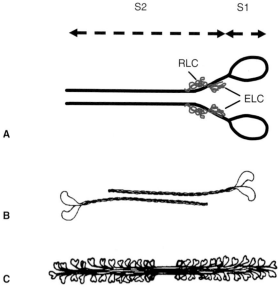

**Figure 1.2** Myosin structure and assembly into thick filaments. **A**, schematic arrangement of myosin subunits showing the S1 and S2 portions of the molecule and the regulatory (RLC) and essential (ELC) light chains. **B & C**, the basic double-headed myosin units aggregating to form a thick filament.

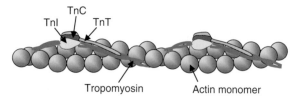

**Figure 1.1** Part of an actin filament together with tropomyosin and troponin (Tn).

or S1 fragment, contains the ATPase activity and is the portion that can combine with actin. The S2 portion includes the flexible region of the molecule and a tail, which combines with other tails, binding the myosin molecules together to form the *thick filaments* (Fig. 1.2B,C). Thick filaments consist of approximately 300 molecules arranged so that the myosin heads are pointing in the opposite directions in the two halves of the filament. Consequently there is a region in the centre of each filament where there are only tails and no projecting heads. This region constitutes about 10% of the total length. Unlike the thin filaments, which vary a little, the thick filaments are very uniform in length throughout the animal kingdom.

In the muscle fibre, thick myosin filaments are arranged so that the thin actin filaments can slide between them (Fig. 1.3). The unit from Z line to Z line is known as a sarcomere and, in mammalian muscle is between 2 and 2.5 μm long when the muscle is held at its natural resting length in the body. Each thick filament is surrounded by six thin filaments, so that each myosin filament may bind to any of six actin filaments (Fig. 1.3). Conversely, each actin filament can interact with three different myosin filaments. At the Z lines

the thin filaments are held in a square array, whereas in the overlap region they are forced into a hexagonal array by the arrangement of the thick filaments. The thin filaments must, therefore, be somewhat flexible and in the I-band region, where there is no overlap, the actin filaments are in transition between the square and hexagonal array and no regular structure is seen.

The nomenclature of the various bands in a sarcomere is shown in Figure 1.4. The A and I bands are so called because of their birefringent properties under the light microscope, the I band being *isotropic* and the A band *anisotropic*. At a more mundane level, the bands can be remembered as being l*i*ght (I) and d*a*rk (A) when seen in longitudinal sections with the electron microscope. The area in the A band where there is no overlap with thin filaments is known as the H zone, in the centre of which is the region of the thick filaments bare of projecting myosin heads. Proteins running across this region give rise to the M line (Fig. 1.4).

Identification of the two contractile proteins and understanding how they are arranged to give the banded appearance of skeletal muscle has been central to understanding of the mechanism of force generation. This knowledge has been derived, first, from observations with the light microscope showing that the A band is of a constant width, while it is the I band that changes as the muscle lengthens and contracts (see Fig. 2.1). Second, the location of the different proteins

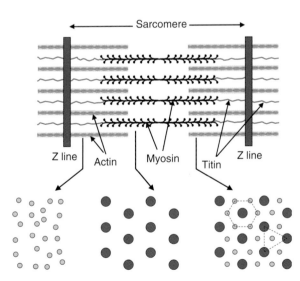

**Figure 1.3** Arrangement of thick and thin filaments to form a sarcomere. Below, cross-sections of the sarcomere where, from left to right, there are only actin filaments, only myosin filaments and, on the right, overlap of actin and myosin.

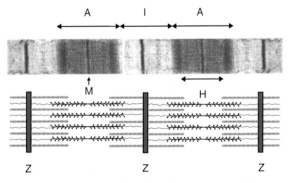

**Figure 1.4** The banded, or striated, appearance of skeletal muscle. **Above**, portion of an electron micrograph of two sarcomeres; **below**, a diagram showing the arrangement of actin and myosin filaments.

was confirmed by experiments in which myosin was extracted from dissociated myofibrils with potassium chloride solution containing ATP or pyrophosphate, showing that, after extraction, the Z line and I bands remained but the A band had been removed. Finally, observations at the electron microscopic level of the way in which myosin monomers in solution aggregate to form the characteristic bidirectional filaments (Fig. 1.2C) indicated the structure and possible function of the thick filaments.

## STRUCTURAL PROTEINS

There are a variety of proteins whose function is to maintain the architecture of the sarcomere. Proteins constituting the M line (Fig. 1.4) keep the myosin filaments in the correct hexagonal arrangement for the actin filaments to slide between.

Titin is an extremely long protein (about 1 μm) which runs from Z line to M line and could be a template on which the myosin monomers condense to make the thick filaments. Titin also helps to provide longitudinal stability to the sarcomere, as it appears to be responsible for the resting force at long sarcomere lengths and prevents the sarcomeres being pulled apart. Nebulin is another very large protein found associated with actin near the Z line, it may strengthen the thin filaments or may act as a template for the actin monomers to form filaments. The Z-line structure contains α-actinin that binds the actin filaments together, while desmin links the Z lines of adjacent myofibrils and serves to keep the Z lines in register.

Dystrophin is another very large protein, the function of which is to anchor the contractile apparatus to the surface membrane and, via connections with other proteins such as dystroglycan, sarcoglycan and laminin, links with the basement membrane (Fig. 1.5). The precise functions of these proteins are not known, but they

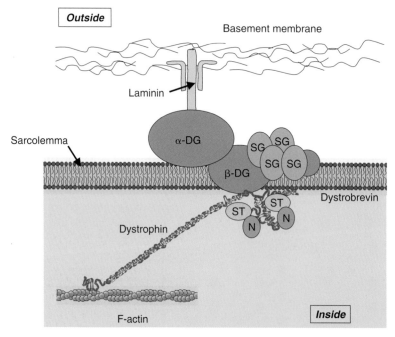

**Figure 1.5** Cytoskeletal proteins connecting the basement membrane and sarcolemma with the underlying contractile structures. DG, dystroglycan; SG, sarcoglycan—glycosylated proteins in the surface membrane. N, nitric oxide synthase (NOS); ST, syntrophin that binds signalling molecules, such as NOS, to dystrophin and dystrobrevin.

are clearly very important as absences or defects are associated with a range of muscle-wasting diseases known as the muscular dystrophies. Duchenne muscular dystrophy and the milder form of Becker dystrophy are associated with an absence, or abnormal form, of dystrophin (see Ch. 15). This cytoskeletal complex is also associated with signalling proteins and nitric oxide synthase, which may be involved in sensing mechanical activity, muscle length etc., leading to functional adaptations.

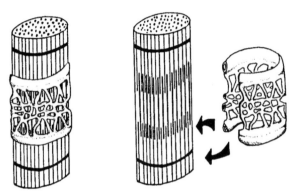

**Figure 1.6** The sarcoplasmic reticulum envelops a myofibril.

## SARCOPLASMIC RETICULUM

Groups of about 200 thick and thin filaments constitute a *myofibril*. Each myofibril is enveloped in a complex membranous bag, known as the sarcoplasmic reticulum, the interior of which is quite separate from the cytoplasm of the fibre (Fig. 1.6). This membrane system is a store for the uptake and release of calcium. The portions lying near the T tubules (see below) are known as the *terminal cisternae*.

## T TUBULAR SYSTEM

The surface or plasma membrane (sometimes also known as the sarcolemma) of the muscle fibre invaginates, forming T tubules which run transversely (hence their name) across the fibre, forming a complex branching network that contacts and, mostly, surrounds every myofibril (Fig. 1.7).

The invaginations of the surface membrane occur twice in every sarcomere approximately at the level of the junction of the A and I bands. In amphibian muscle, in contrast to mammalian muscle, there is only one T tubule per sarcomere, and this runs at the level of the Z line. Where the T tubules meet the sarcoplasmic reticulum, the

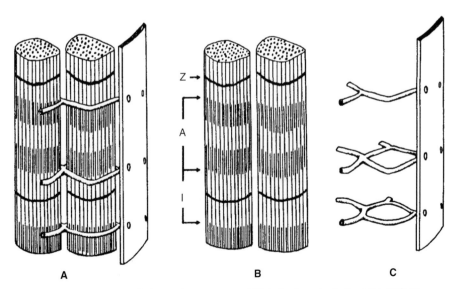

A            B            C

**Figure 1.7** T tubules in relation to the sarcomere. T tubules invaginate from the surface membrane of the muscle fibre (**C**) and contact the myofibrils at the junction of the A and I band (**A** & **B**).

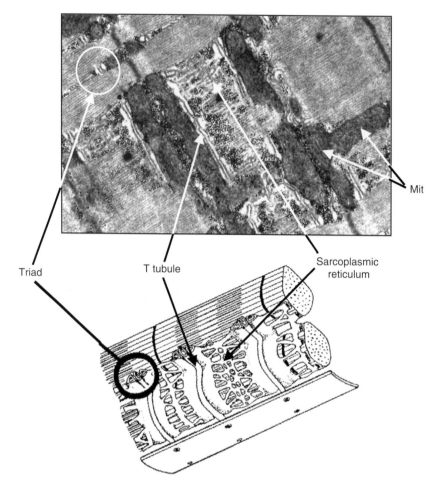

**Figure 1.8** T tubules, sarcoplasmic reticulum and the myofibrils. Structure of a triad is shown within the ring. Mit, mitochondria.

two membranes run very close to one another, and with the electron microscope dense 'feet' can be seen linking the two sets of membranes.

In electron micrographic sections, the T tubules are often seen cut in cross-section with a portion of the sarcoplasmic reticulum on either side (Fig. 1.8); this is known as a 'triad'. When associated with sarcoplasmic reticulum in a triad, the T tubule is flattened; otherwise it tends to be circular in cross-section.

## THE MYOFIBRIL IN RELATION TO THE FIBRE AND MUSCLE

Bundles of 100 to 400 filaments form *myofibrils*, which are separated from adjacent myofibrils by sarcoplasmic reticulum, T tubules and sometimes mitochondria. Myofibrils vary in size but average around 1 μm in diameter, making up about 80% of the volume of a muscle fibre. The numbers of myofibrils vary with the size of the muscle fibre and there may be as few as 50 in a developing fetal muscle fibre. There are about 2000 myofibrils in an adult muscle fibre. Bundles of muscle fibres are then arranged together to form the anatomical muscle (Fig. 1.9).

## MUSCLE ORGANIZATION

Each muscle fibre is bounded by its sarcolemma or plasma membrane. At resting lengths the membrane is folded with small indentations or

Myofibril

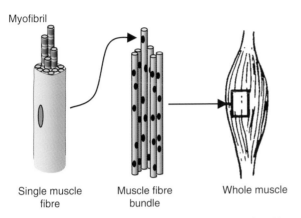

Single muscle          Muscle fibre          Whole muscle
fibre                  bundle

**Figure 1.9** Diagrammatic representation of the relationship between myofibrils, muscle fibres and the whole anatomical muscle. In reality the myofibrils are very much smaller, in relation to the muscle fibre, than shown here.

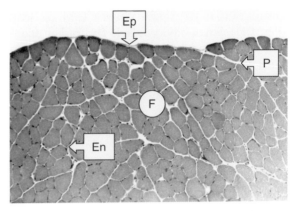

**Figure 1.11** Transverse section through a portion of mouse skeletal muscle (light microscopy, haematoxylin and eosin stain). F, fascicle; En, endomysium; P, perimysium; Ep, epimysium.

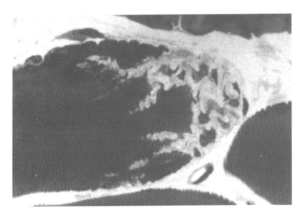

**Figure 1.10** Low-power electron micrograph of a myotendinous junction.

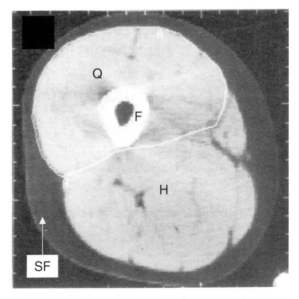

**Figure 1.12** Computed tomogram of a cross-section through the human mid-thigh, showing the quadriceps (Q) and hamstring (H) muscles together with the femur (F) and subcutaneous fat (SF).

*caveolae*, which smooth out as the muscle is stretched and are probably important in allowing considerable change in muscle fibre length without causing damage to the surface membrane. Outside the plasma membrane is the basement membrane, which is not a membrane in the usually accepted sense of the word in that it does not have a lipid bilayer structure but is composed of a loose glycoprotein and collagen network. It is freely permeable and may surround more than one fibre. Following muscle fibre damage, the basement membrane forms a framework within which regeneration occurs.

At the ends of a muscle fibre the outer membranes become irregular and indented to form a

close link with the connective tissue (Fig. 1.10). The connective tissue elements all come together to form the tendons, which join the muscle to the bony skeleton.

Between the muscle fibres are fibroblasts, which secrete collagen fibres to form a thin connective tissue matrix, the endomysium. A thicker layer of connective tissue, the perimysium, surrounds groups of 10 to 100 muscle fibres to form fascicles (Fig. 1.11). Small blood vessels and motor axons traverse the perimysial spaces to make connections with the muscle fibres. Muscle spindles are also found enclosed in connective tissue envelopes in the perimysium. A thick outer connective tissue layer, the epimysium, covers each muscle. Fibroblasts, and their major product collagen, have an important role maintaining

both the structure of a muscle and an environment within which the contractile muscle fibres can function.

In the average quadriceps muscle there are about one million fibres in a section taken at the mid-thigh level (Fig. 1.12); however, the muscle has a complex structure so that a single cross-section will not include all the fibres, and the whole muscle may consist of several million.

REFERENCES AND FURTHER READING

Hanson J, Huxley H E 1955 The structural basis of contraction in striated muscle. Symposia of the Society for Experimental Biology IX: 228–264

Huxley A F 1976 Looking back on muscle. In: The pursuit of nature, pp 23–64. Cambridge: Cambridge University Press

Knight P, Trinick J 1987 The myosin molecule. In: Squire J M, Vibert P J (eds) Fibrous protein structure, pp 247–281. London: Academic Press

Needham D 1971 Machina carnis. Cambridge: Cambridge University Press

Offer G 1987 Myosin filaments. In: Squire J M, Vibert P J (eds) Fibrous protein structure, pp 307–356. London: Academic Press

Page S G, Huxley H E 1963 Filament lengths in striated muscle. Journal of Cell Biology 19: 369–390

Peachey L D 1965 The sarcoplasmic reticulum and transverse tubules of the frog's sartorius. Journal of Cell Biology 25(Suppl 3): 209–231

Rayment I, Holden H M, Whittaker M et al 1993a Structure of the actin–myosin complex and its implications for muscle contraction. Science 261: 58–65

Rayment I, Rypniewski W R, Schmidt-Base K et al 1993b Three-dimensional structure of myosin subfragment—1: A molecular motor. Science 261: 50–58

Squire J M, Luther P K, Trinick J 1987 Muscle myofibril architecture. In: Squire J M, Vibert P J (eds) Fibrous protein structure, pp 423–450. London: Academic Press

## QUESTIONS TO THINK ABOUT

1. What are the different components that make up the thin filaments in a sarcomere?

2. What are the different components that make up thick filaments?

3. What is the difference between a muscle fibre and a myofibril? What separates one myofibril from the next?

4. Why do thin filaments form a hexagonal array in the overlap zone of the sarcomere?

5. Where is titin found within the sarcomere and what is its possible role?

6. How is the contractile apparatus connected with the surface and basement membranes of the muscle fibre and does this have any functional significance?

7. What is a triad and from which membrane system are the T tubules derived?

8. The area of the surface membrane of a muscle fibre is much greater than would be predicted from treating the fibre as a cylinder. Why is this?

9. What is the function of the sarcoplasmic reticulum?

10. What differences in structure are there between fast and slow fibres, viewed at the level of the electron micrograph?

# 2

# Mechanisms of force generation

Our understanding of how muscle proteins generate force and movement has come about by elucidating the structure of the actin and myosin filaments and the way they are organized in the sarcomere, and then from theoretical models showing how these structures can account for the behaviour of contracting muscle. Although details of cross-bridge interactions continue to be discovered, the major advance in understanding muscle contraction occurred in the 1950s and 1970s, largely from the work of A. F. Huxley and H. E. Huxley—unrelated and working separately—but also owing much to observations made by A. V. Hill and others in the 1930s, as well as the work of microscopists dating back to the previous century.

In this chapter the mechanical properties of contracting muscle are presented together with a discussion of the evidence that these properties provide about the nature of the mechanism generating force in skeletal muscle.

## NATURE OF THE INTERACTION BETWEEN THICK AND THIN FILAMENTS

There are two properties of muscle that provide important clues about the nature of the contractile apparatus. The first is the way that force varies with the length of the muscle, and the second is the characteristic relationship between the force that can be developed and the velocity of movement, during either shortening or lengthening.

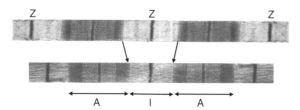

**Figure 2.1** Changes in the I band, but not the A band, with shortening. **Above**, sarcomeres fixed at a relatively long length. **Below**, sarcomeres fixed at a shorter length. Note that the I band has become shorter but the A bands remain the same length.

## Length–force relationship and the sliding filament theory

It had been known since the nineteenth century that, as a muscle contracts, the A band remains of constant length while the I band becomes shorter (Fig. 2.1). The implication of this is that, while the thick myosin filaments remain fixed, the thin actin filaments slide in and out. The increasing darker region that can be seen in the A band indicates a region of filament overlap. It was suggested, therefore, that force is developed by the interaction of actin and myosin filaments as they overlap, and this became known as the *sliding filament theory*. Proof was provided by careful comparison of the extent of filament overlap and the force generated by the muscle. If the theory is correct then force should be proportional to the extent of overlap, which can be estimated knowing the overall sarcomere length and the length of the thick and thin filaments. Figure 2.2A shows the general shape of the force–length relationship of skeletal muscle. Force is low at short muscle lengths, rises to a maximum, and then falls as the muscle is stretched further. By extrapolating the line at longer lengths, a value can be predicted at which no force is developed; this corresponds to the length at which the actin and myosin filaments no longer overlap (I in Fig. 2.2B). The careful measurements made by Gordon et al (1966) show that on the right side of the relationship the linear decline in force is, indeed, proportional to the degree of filament overlap (Fig. 2.2B).

The reason why force decreases relatively rapidly on the left-hand side of the relationship

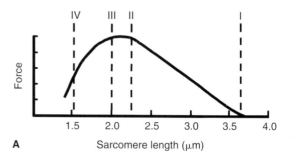

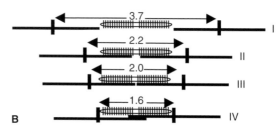

**Figure 2.2** Isometric force at different sarcomere lengths. **A**, force generated; **B**, arrangement of filaments at different lengths. At lengths less than 2.0 μm thin filaments begin to overlap, and at still shorter lengths the thick filaments come into contact with the Z lines. Values are for frog muscle (data from Gordon et al 1966); mammalian thin filaments are slightly longer, so the corresponding sarcomere lengths are: I, 4.0 μm; II, 2.5 μm; III, 2.4 μm; and IV, 1.6 μm.

(Fig. 2.2A) is that at very short lengths the thin filaments start to overlap one another in the centre of the sarcomere and the thick filaments come into contact with the Z lines (IV in Fig. 2.2B).

Although the interaction of thin and thick filaments is still referred to as a *theory*, in reality it might more properly be called the sliding filament *mechanism* of muscle contraction.

## The interaction between actin and myosin

The sliding filament mechanism shows how force is proportional to the extent of overlap of the actin and myosin filaments, but it does not specify the type of interaction between the two filaments that generates force. Again, with hindsight, it is easy to see that the greater the overlap the larger the number of myosin heads that can react with actin, and that the force generated is the sum of all the small forces produced by individual cross-bridges, but this was not always so obvious and there have been suggestions that when actin and myosin interact they form the equivalent of a coiled spring that generates force. However, contracting muscle does not behave as though it were made of a single spring.

If an active muscle is rapidly shortened by as little as 1% of its total length, the force momentarily drops to near zero before redeveloping. If the actin and myosin had formed into some kind of long spring when the muscle was activated, the force would be expected to decrease in proportion to the change in its overall length, that is, it would obey Hooke's law. A 1% decrease in length for a long spring would be expected to drop the force by 1%, not the value close to 100% that is observed. This important observation suggests that force is generated by components that are very short, so that a change of 1% in the sarcomere length is sufficient to take all the stretch out of them. This fits well with the notion that force is generated by numerous cross-bridges formed by the myosin head regions, which are each active over a small distance.

## MUSCLE SIZE AND FORCE

The isometric force produced by a muscle is proportional to the cross-sectional area of the muscle, rather than its length. Two extreme ways of arranging the same amount of contractile material are shown in Figure 2.3.

Working outwards from the central Z line of a muscle fibre with the sarcomeres arranged in series, it can be seen (Fig. 2.3A) that the forces exerted by each half-sarcomere on the adjacent Z line are opposed to one another and so do not summate along the length of the fibre. The intermediate sarcomeres serve to form a rigid connection between the two ends and, for the purposes of generating isometric force, could be replaced with a piece of string.

The arrangement in Figure 2.3B, in contrast, will develop a much greater force. Although the generation of isometric force is an important function of muscle, it is also vital that the muscle should be able to shorten. During a contraction

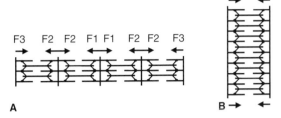

**Figure 2.3**  Force generated by sarcomeres in series and in parallel. **A**, sarcomeres in series. The forces F1 and F2 are opposed, leaving only F3 to exert force at the ends of the muscle. **B**, the same number of actin and myosin filaments arranged in parallel to give four times the isometric force of **A**.

each sarcomere will shorten by a similar amount, say 1 μm. If there are 1000 sarcomeres in series, the whole muscle will shorten by 1 mm; if there are 10 000 sarcomeres in series, the muscle will shorten by 1 cm, etc.

The distance a muscle can shorten depends on its length. In general, muscles work over a range where they shorten by about half their initial length, so a muscle such as that illustrated in Figure 2.3B may be able to exert a very high isometric force and be useful in stabilizing joints, but would not be much use for bending the arm or flexing the knee—movements that require muscles to shorten over appreciable distances. In general, short muscles are good for stabilizing the body; they can generate high forces for the amount of contractile material they contain, but cannot shorten very far or very fast. Long muscle, on the other hand, can shorten over long distances, enabling large movements of the limbs, and they can do so relatively fast. However, in terms of the total amount of contractile material, they generate relatively low forces.

## FORCE–VELOCITY CHARACTERISTICS

We have seen that the force generated by a muscle depends on the filament overlap, and therefore sarcomere length, but the force generated also varies with the velocity at which it is shortening.

As the velocity of muscle shortening increases, so the force sustained by the muscle rapidly diminishes (Fig. 2.4), eventually reaching a velocity at which force can no longer be sustained at all; this is the maximum velocity of shortening ($V_{max}$). The force at zero velocity of shortening is the isometric force (referred to as $F_0$). Expressing the force ($F$) at a particular velocity as a fraction of $F_0$, it is possible to compare muscles of different sizes and therefore different isometric strengths. The basic hyperbolic form of this relationship was first described in the 1920s and 1930s when there was a surge of interest in the relationships between force, velocity and energy liberation during contraction. During an isometric contraction there is no movement, so the muscle does no external work and all the energy liberated appears in the form of heat. During shortening, heat is still produced but the muscle also performs work, which is the product of the force and distance moved. It was first observed by Fenn in 1923 and 1924 that during shortening the total energy liberated, in the form of heat plus work, is greater than during an isometric contraction (Fenn effect). This observation stimulated a great deal of work and speculation about the nature of the contractile process and led A. V. Hill in 1938 to propose an equation describing the shape of the force–velocity relationship:

$$(F + a)V = b(F_0 - F),$$

although Hill himself used $P$ and $P_0$ to represent force. This is the equation of a hyperbola in which the axes have been shifted by the constants $a$ and $b$. The equation is a valuable way of describing the characteristics of a muscle and of estimating $V_{max}$ where this cannot be easily measured.

Although isometric force is independent of the number of sarcomeres in series, this is not the case for speed of shortening. At the onset of contraction, all sarcomeres in a muscle fibre will begin to shorten at the same time and at the same velocity. If the muscle were only one sarcomere long and shortened from 3 to $2\,\mu m$ in one-tenth of a second then the velocity of shortening would be $10\,\mu m\,s^{-1}$. If the muscle consisted of 100 sarcomeres in series, the shortening velocity would be $1\,mm\,s^{-1}$, and for a muscle an inch long (2.5 cm) containing about 10 000 sarcomeres in series, the velocity of shortening would be $10\,cm\,s^{-1}$. To compare speed of shortening in muscles of different lengths, the velocity is often expressed as muscle lengths per second or as sarcomere lengths per second. Mammalian muscles vary greatly in their speed of shortening, but human muscles have a maximum velocity of shortening of around 4 muscle lengths per second.

Power is the product of force and velocity (Fig. 2.5). As force is proportional to the cross-sectional area of a muscle, and velocity to the length, it follows that power is proportional to the product of these, namely volume. Thus, a short fat muscle (e.g. Fig. 2.3B) will generate a high force but have a low maximum velocity of shortening, while a long thin muscle (e.g. Fig. 2.3A) will produce little force but shorten rapidly (Fig. 2.5). The muscles

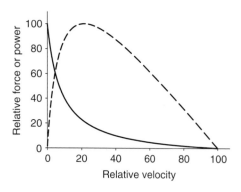

**Figure 2.4** Example of a force–velocity (solid line) and a power–velocity (dashed line) relationship.

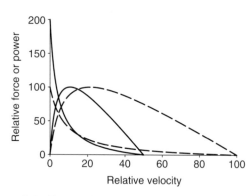

**Figure 2.5** Force–velocity and power–velocity relationships in muscles with different lengths and cross-sectional areas. Solid line, short fat muscle; dashed line, long thin muscle.

will, however, have the same volumes and, therefore, maximum power outputs. Maximum power is usually obtained at about one-third $V_{max}$ so that, although the maximum power may be the same in the two muscles, the velocity at which this occurs will be different (Fig. 2.5).

Hill's characteristic equation is an empirical description of the experimental data and does not embody any hypothesis about the way in which force is generated; it is not immediately obvious why force varies in this way as the muscle shortens. In contrast to muscle, when an elastic band is allowed to shorten over a fixed distance it will sustain a certain average force, and this force will be the same no matter what the speed of shortening. This behaviour is quite different to the way muscle functions. Resolution of this question had to await the development of models of contraction based on what is known of the interaction of the myosin cross-bridges.

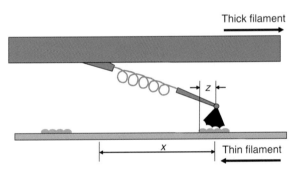

**Figure 2.6**  Rotating head model. The head of the cross-bridge can bind to actin in a number of stable configurations at different angles and therefore with different values of z. The length of, and thus the force in, the spring in the cross-bridge is dependent on the two variables, x and z. The arrows indicate the direction of relative movement of the thick and thin filaments. (Redrawn from the data of Huxley & Simmons 1971.)

# CROSS-BRIDGE KINETICS

The current ideas of cross-bridge action and the model of the myosin molecule with a head that rotates and stretches a compliant portion is largely the result of the theories put forward by A. F. Huxley in 1957 and extended by Huxley & Simmons in 1971.

Models of cross-bridge action have evolved to accommodate various features of muscle function. The three-state model was introduced in 1971 by Huxley & Simmons and incorporates a flexible head so that the extension of the compliant portion of the myosin molecule is determined by both the displacement of the actin binding site from the equilibrium position $(x)$, and also the rotation of the head $(z)$ (Fig. 2.6).

It is suggested that the head can bind to the actin and that it is the subsequent rotation of the head that stretches the compliant S2 portion of the myosin molecule and generates force (Fig. 2.7). However, transitions between the bound states are very rapid, so that for movements other than very rapid transients the model can be considered as a two-state model in which the cross-bridges are either attached or detached (Fig. 2.8).

## Isometric force

The maximum number of cross-bridges that can be formed will be set by the degree of filament overlap and the number of actin binding sites exposed by the binding of calcium to troponin. However, only a proportion of these will be attached at any one moment and this fraction will be a function of the rates of attachment and detachment. The rate constant for attachment is usually designated $f$ and that for detachment $g$ (Fig. 2.8).

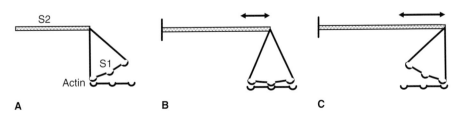

**Figure 2.7**  Cross-bridge attachment during isometric contractions. Attachment and rotation of the myosin head (**A–C**) causes extension of the compliant S2 portion. Double-headed arrows indicate the amount of stretch and thus force in S2.

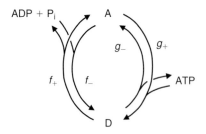

**Figure 2.8** Huxley's theory (1957) postulates the existence of two cross-bridge states: attached (A) and detached (D). The cycle of attachment and detachment is driven by ATP splitting. $f_+$ and $g_+$ are the rate constants for the forward (clockwise) processes; $f_-$ and $g_-$ are those for the reverse reactions. $P_i$, inorganic phosphate.

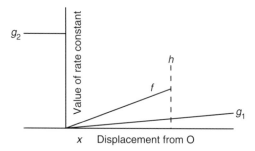

**Figure 2.9** Rate constants for attachment ($f$) and detachment ($g_1$ and $g_2$) of myosin to actin as a function of cross-bridge displacement from the equilibrium position O. $h$ is the maximum displacement of the myosin head at which binding can occur. $f$ is zero for negative values of $x$, and $g_2$ has a high and constant value.

If $n$ is the fraction of cross-bridges attached, then the rate of attachment is given by $f \times (1 - n)$, and the rate of detachment by $g \times n$. During a strictly isometric contraction when force has reached a constant value, the rates of attachment and detachment will be equal, so:

$$f(1 - n) = g \cdot n$$

or

$$n = f/(f + g)$$

If the rate constants for attachment and detachment were equal, the proportion of attached cross-bridges would be 50%. During an isometric contraction the proportion of attached cross-bridges is thought to be about 80%, which means that $f$ is about four times $g$. The proportion of attached cross-bridges, given by the fraction $n$, is independent of the absolute values of $f$ and $g$ provided there is sufficient time for equilibrium to be established between attached and detached cross-bridge states. This will only truly be the case during isometric contractions when the actin and myosin binding sites remain in the same relative positions.

To explain the variations in force generated at different velocities of contraction it is necessary to consider how the rates of attachment and detachment may vary with the displacement of the individual cross-bridge. Figure 2.9 shows the way in which Huxley (1957) suggested $f$ and $g$ vary with cross-bridge displacement from the equilibrium position (O), where there is no extension in the compliant S2 portion and no force generated. The displacement from this position is $x$ and the maximum displacement at which attachment can occur is $h$. The rate constants for attachment ($f$) and detachment ($g_1$, for positive values of $x$) are postulated to vary linearly with displacement for positive values of $x$. Attachment is not considered possible for negative values of $x$, and the constant for detachment in this region is assigned a constant value ($g_2$).

During shortening, cross-bridges that are carried beyond the central position will generate force opposing the movement. To keep this opposing force to a minimum it is necessary that there should be a rapid dissociation once the cross-bridge moves into regions where the value of $x$ is negative. This is why a high value is given to $g_2$.

## Force during shortening

During contractions where the muscle is shortening, the force sustained ($F$) is less than isometric force ($F_0$) for three reasons. First, the faster the movement the fewer cross-bridges that will be attached. As the actin binding sites move past the myosin cross-bridges, there will be only a limited time during which attachment can take place. The faster the velocity of shortening, the shorter the time during which the myosin head is in the vicinity of an actin binding site and consequently the smaller the proportion of cross-bridges that will manage to attach. Second, those cross-bridges that are attached will, on average, be less stretched and thus generate less force. Third,

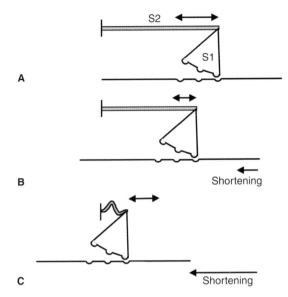

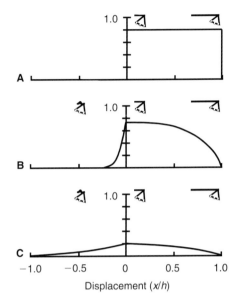

**Figure 2.10** Cross-bridge attachment during shortening. **A**, isometric condition, S2 fully extended. **B**, slow shortening with the actin filament moving from right to left; the average extension of S2 is less than in **A**. **C**, rapid shortening with the S2 portion in compression. Double-headed arrows indicate the amount of stretch or compression of S2.

**Figure 2.11** Distribution of cross-bridges calculated from Huxley's 1957 theory. **A**, isometric contraction; **B** & **C**, increasing velocities of shortening. Vertical axis, the proportion of available cross-bridges; horizontal axis, displacement of myosin cross-bridge from the equilibrium position in units of $h$ (see Figs 2.6 & 2.9).

a proportion of cross-bridges will be carried into positions where they oppose movement (i.e. negative values of $x$) (Fig. 2.9) and thereby reduce the total force. The latter two situations are illustrated in Figure 2.10. As the velocity of shortening increases there will come a time when the force generated by the cross-bridges attached in regions with positive values of $x$ will be matched by the opposing forces of cross-bridges that have not detached and have been carried into negative values of $x$. At this velocity, the forces cancel each other out; this is known as the maximum velocity of unloaded shortening, or $V_{max}$.

The linear manner in which the rate constants were chosen to vary with displacement from the equilibrium position in Huxley's 1957 model (Fig. 2.10) is arbitrary but has the advantage that it generates soluble differential equations. Using these equations the likelihood of a cross-bridge in a certain position being attached for a given velocity of shortening can be calculated. Thus the proportion of cross-bridges that are attached can be calculated as a function of

displacement from the equilibrium position as the actin and myosin filaments slide past one another.

Figure 2.11 shows the distribution of cross-bridges for different types of contraction. In an isometric contraction (A), cross-bridges are all in the range O to $h$. In (B), with slow sliding of the filaments past one another, fewer cross-bridges are attached near to the position $h$ because none can be formed at lengths greater than $h$ and those that are formed are moved away towards the centre. Midway between $h$ and O, the number attached is not much reduced because those that are carried away towards O are replaced by cross-bridges carried away from the region closer to $h$. Some cross-bridges formed close to O will have been carried into the region beyond the equilibrium position so that the bridges are now compressed and oppose movement. At a high velocity of shortening (C), there is little time for cross-bridge attachment so that the total number of attached bridges is reduced. A large number of those that do manage to attach are carried beyond the equilibrium position and therefore oppose force generation.

The force generated by an individual cross-bridge depends on the extension of S2, which in turn depends on the displacement of the binding site relative to the equilibrium position O. The total force generated is therefore proportional to the integral of $n \times x$, that is, the number of cross-bridges attached in each position multiplied by their displacement from the equilibrium position.

When the opposing force due to cross-bridges in compression equals the force generated by the cross-bridges in the region O to $h$, the net result is no force production by the muscle; this occurs at the maximum velocity of shortening. This velocity is critically dependent on the value of $g_2$, because the higher the value of $g_2$ the fewer the number of cross-bridges that will remain in the region where they are compressed and oppose movement and reduce the overall force generated by the muscle.

See Woledge et al (1985) and Gordon et al (2000) for general reviews of cross-bridge kinetics.

## Force during stretch

In the body, muscles work in pairs, while one muscle shortens its antagonist is stretched during movements such as lowering weights. When walking down stairs the quadriceps is extended in lowering the bodyweight while the calf muscle can also be stretched when landing on a pointed foot as the heel comes down to the ground (Fig. 2.12A).

The force generated during this type of movement is considerably greater than the isometric force (Fig. 2.12B) and varies with velocity. With increasing velocity of stretch, force begins to plateau, reaching a value of about 1.8 times the isometric force (Fig. 2.13) (Katz 1939, Lombardi & Piazzesi 1990).

The increased force can be explained in terms of Huxley's model in that during a stretch the compliant S2 portions of individual cross-bridges are stretched further than is normally the case during isometric contractions (Fig. 2.14A). During relatively slow stretches, a proportion of cross-bridges will still go through the full cycle of attachment,

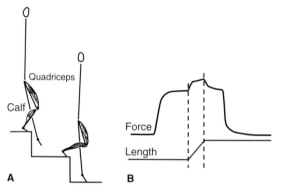

**Figure 2.12** Stretching muscles. **A**, the quadriceps are stretched when lowering the bodyweight down a step. **B**, mouse soleus muscle stimulated to develop maximum isometric force and then stretched at a constant rate.

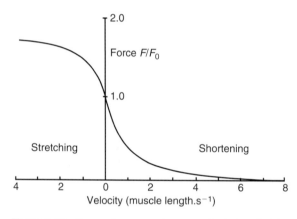

**Figure 2.13** Force of mouse soleus muscle during stretch and shortening at different velocities, 26°C.

rotation of the myosin head and detachment. However, as the velocity of stretch increases, fewer cross-bridges will be able to complete the cycle and an increasing proportion of myosin heads will remain in position (ii) (Fig. 2.14A). When these cross-bridges are stretched even further, they will be forcibly detached. It is suggested that these myosin heads are able to reattach to actin very rapidly, giving a sustained high force during long stretches at relatively low energetic cost.

Figure 2.14B shows the distribution of cross-bridges as predicted by Huxley's sliding filament model assuming that the rate constant of detachment ($g_1$) continues to increase linearly beyond $h$.

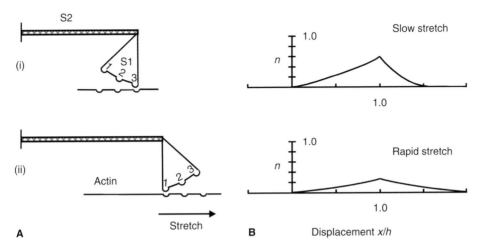

**Figure 2.14** Behaviour of cross-bridges during stretch. **A**, movement of actin from left to right produces greater stretch of the S2 component than in the isometric condition. Note that for slow stretches (i) some myosin heads can reach position 3 and complete the cycle, but for fast stretches (ii) most will remain in position 1. **B**, distribution of cross-bridges for slow and rapid stretch.

With increasing speed of stretch, the number of cross-bridges will decrease but those attached will sustain more force, thus the force maintained by a muscle during stretch tends towards a plateau at higher velocities.

The overall shape of the curve shown in Figure 2.13, with force increasing with stretch and decreasing with shortening, has important consequences for the stability of sarcomeres acting in series. If one sarcomere were stronger than the next, the stronger might pull out and extend the weaker, eventually destroying the fibre. However, when the sarcomeres begin to move, the force generated by the stronger will fall as it shortens (moving down the right-hand portion of the curve in Fig. 2.13), while the force of the weaker sarcomere will increase as it is stretched (moving up the left-hand side of Fig. 2.13). This may be one mechanism whereby sarcomeres of unequal strength (due to differences in length, activation or even damage) can coexist and function to transmit force along the length of the fibre.

## BIOCHEMISTRY OF FORCE GENERATION

There is general agreement that the interaction of actin, myosin and ATP proceeds in a stepwise fashion and that the intermediate stages can, or eventually will, be identified with different mechanical steps in the cross-bridge cycle.

The first indication of the complex nature of the actomyosin ATPase was the observation that, on adding actin to a mixture of myosin and ATP, there is an initial rapid appearance of phosphate (phosphate burst) before the system settles down to a steady rate of ATP hydrolysis. The explanation for the phosphate burst is that if myosin is first incubated with ATP the hydrolysis products, ADP and $P_i$, remain bound to the protein and are released only when actin and myosin combine.

Detailed kinetic studies have demonstrated a large number of possible intermediates. The main steps thought to be involved in a normal cross-bridge cycle are shown in Figure 2.15A, and their probable relation to the mechanical events is indicated in Figure 2.15B.

Attachment of actin and myosin (i) is a reversible process that gives stiffness to the muscle (i.e. it will resist if stretched), but does not itself generate force. The release of phosphate from the actomyosin complex (ii) is thought to initiate the changes that result in force generation (rotation of the S1 head in this model). Towards the end of the rotation phase ADP is released (iii) and the actomyosin complex can then bind ATP (iv). Having done so, the actin and myosin dissociate with the ATP bound to myosin (v). The bound ATP is then

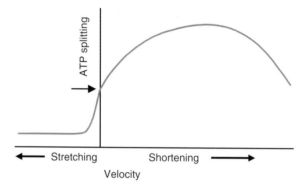

**Figure 2.16** ATP turnover in muscle allowed to shorten or while being stretched at different velocities. Arrow indicates ATP turnover during an isometric tetanus.

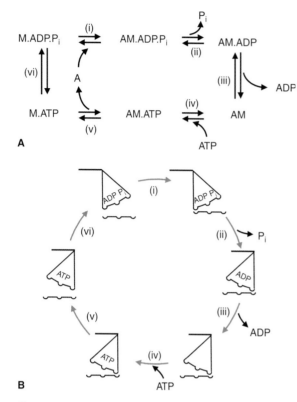

**Figure 2.15** Stages in the cross-bridge cycle, corresponding to the different biochemical steps (**A**) and showing the probable functional changes (**B**). Force is generated between steps (i) and (iii). AM, actomyosin; M, myosin, $P_i$, inorganic phosphate.

hydrolysed and the products remain bound to the protein (vi); this last process is thought to activate the S1 unit, making it ready to bind to actin again. All the reactions in Figure 2.15A are shown as reversible, but in practice some are more reversible than others. Reaction (ii), the release of $P_i$, may be quite sensitive to the phosphate concentration, but the back reaction of (vi), the hydrolysis of ATP and the activation of the myosin head, is unlikely to be energetically favourable.

## ATP splitting during contraction

With a judicious selection of rate constants, the Huxley model produces a remarkably good fit of the observed force–velocity characteristics for skeletal muscle. It also gives an explanation of the Fenn effect: if one molecule of ATP is hydrolysed

with every cross-bridge dissociation, the rate at which heat and work are produced will depend on the rate of cross-bridge turnover. In the isometric state, where cross-bridges are in positions between O and $h$, the rate constant for cross-bridge detachment is relatively low. There will be a modest rate of cross-bridge cycling and ATP turnover. However, in the region beyond the equilibrium position where attached cross-bridges oppose movement (negative values of $x$), the rate of detachment $g_2$ is high and therefore turnover and liberation of energy will be high.

As the velocity of shortening increases, so the number of cross-bridges carried beyond the equilibrium position increases, accounting for the greater production of heat plus work. The rate of energy liberation will, however, begin to level off with increasing velocity because, although turning over more rapidly, there will be fewer cross-bridges actually attached.

On the basis of one molecule of ATP split for every cross-bridge detachment, the model predicts that ATP splitting would increase with increasing velocity both of shortening and stretching. However, heat production falls to low values when muscle is stretched, a fact first noticed by Fenn (1923). Later, Curtin & Davies (1973) showed that ATP splitting was likewise very low during stretching (Fig. 2.16).

To accommodate this observation, the model requires only minor modification, namely the existence of an attached cross-bridge state where detachment does not involve ATP splitting—this could be the back reaction of *f*. Detachment during shortening would mainly involve ATP splitting but, during stretch, detachment would be primarily a mechanical process. Thus, dissociation when the myosin head has fully rotated would require ATP in reaction (iv) in Figure 2.15, whereas forcible detachment before this stage (for example, (ii) in Figure 2.14A) would not.

REFERENCES AND FURTHER READING

Curtin N A, Davies R E 1973 Chemical and mechanical change during stretching of activated frog skeletal muscle. Symposia on Quantitative Biology XXVII: 619–626

Fenn W O 1923 A quantitative comparison between the energy liberated and the work performed by the isolated sartorius muscle of the frog. Journal of Physiology 58: 175–203

Gasser H S, Hill A V 1924 The dynamics of muscular contraction. Proceedings of the Royal Society B 96: 398–437

Gordon A M, Huxley A F, Julian F J 1966 The variation in isometric tension with sarcomere length in vertebrate muscle fibres. Journal of Physiology 184: 170–192

Gordon A M, Homsher E, Regnier M 2000 Regulation of contraction in striated muscle. Physiological Reviews 80: 853–924

Hill A V 1938 The heat of shortening and the dynamic constants of muscle. Proceedings of the Royal Society B 126: 136–195

Huxley A F 1957 Muscle structure and theories of contraction. Progress in Biophysics and Biophysical Chemistry 7: 255–318

Huxley A F, Simmons R M 1971 Proposed mechanism of force generation in striated muscle. Nature 233: 533–538

Huxley H E 1969 The mechanism of muscular contraction. Science 164: 1356–1366

Katz B 1939 The relation between force and speed in muscular contraction. Journal of Physiology 96: 45–64

Lombardi V, Piazzesi G 1990 The contractile response during steady lengthening of stimulated frog muscle fibres. Journal of Physiology 431: 141–171

Woledge R C, Curtin N A, Homsher E 1985 Energetic aspects of muscle contraction. Monographs of the Physiological Society, no. 41. London: Academic Press

## QUESTIONS TO THINK ABOUT

1. Beyond optimum length, force decreases in a linear fashion, but why is the decrease not linear below optimum length?

2. If every cross-bridge cycle uses the same amount of energy, would you expect the energy costs of isometric contractions at different lengths to be proportional to the force developed?

3. What happens with the length–force relationship of a parallel fibred muscle if:
   a. the muscle becomes twice as thick?
   b. the muscle becomes twice as long?
   What happens to the total energy required to give the same isometric force in these two situations?

4. Explain why the peak power of a muscle is the same if it becomes twice as long and twice as thin. What happens to the velocity at which peak power is obtained?

5. In what ways does the performance of a muscle that is shortening differ from that of a stretched rubber band?

6. Explain why force generated during a shortening of a muscle is less than that during an isometric contraction.

7. Do all attached cross-bridges exert the same force?

8. What would happen if the ATP of a contracting muscle were suddenly depleted?

9. What are eccentric muscle contractions and what physiological roles do they play in normal movement and activity?

10. Explain the shape of the eccentric part of the force–velocity relationship in terms of the Huxley model.

# 3

# Control of the contractile process

For voluntary skeletal muscle, the control of force has to be accurate and precise with the contractile machinery being switched on and off rapidly to allow for complex coordinated movements. This chapter deals with the way in which muscle responds to electrical stimulation.

## THE NEUROMUSCULAR JUNCTION AND POSTSYNAPTIC MEMBRANE

A neuromuscular junction is the synaptic connection at which the axon branch of the motoneuron meets the muscle fibre. Neuromuscular junctions are not often seen in EM preparations from biopsy material as there is only one junction per fibre and the chances of sectioning this region are small. The appearance is of very irregular infoldings of the muscle plasma membrane (sarcolemma), and there are often nuclei and accumulations of glycogen and mitochondria just beneath the junction (Fig. 3.1).

In the presynaptic terminal (A in Fig. 3.1), acetylcholine (ACh) is stored in vesicles, together with ATP and the peptide hormone calcitonin gene-related peptide (CGRP). Action potentials, arriving at the axon terminal, open voltage-sensitive calcium channels and the influx of calcium causes the synaptic vesicles to fuse with the presynaptic nerve membrane and release their contents into the synaptic cleft. The release is, therefore, dependent on the presence of external calcium, and release is depressed by high magnesium concentrations. The amount of ACh released from a single synaptic vesicle is referred to as a quantum.

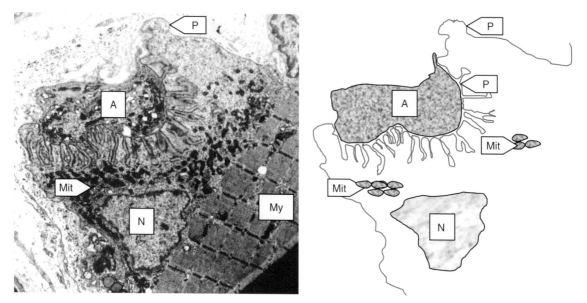

**Figure 3.1** Neuromuscular junction. A, nerve axon; N, nucleus; My, myofibrils; P, muscle fibre plasma membrane; Mit, muscle mitochondria.

The postsynaptic muscle fibre membrane contains acetylcholine receptors (AChRs), which are situated at the crests of the folds. Cholinesterase, the enzyme that hydrolyses acetylcholine, is synthesized by the muscle fibre and secreted into the synaptic cleft where it binds to the basement membrane, which fills the cleft.

Binding of acetylcholine to the postsynaptic receptor causes a depolarization of the muscle fibre membrane. The extent of the depolarization depends on the number of receptors binding ACh, which in turn depends on the number of synaptic vesicles that have discharged into the synaptic cleft.

A single quantum of ACh produces a depolarization of the muscle membrane known as a *miniature end-plate potential* (MEPP). If sufficient quanta of ACh are released at the same time, the MEPPs will summate and produce a large enough depolarization to initiate an action potential (Fig. 3.2). The postsynaptic muscle membrane is rich in $Na^+$ channels, particularly in the folds of the muscle membrane, and this ensures that depolarization leads to a large action potential, which will propagate along the surface and T tubular membranes of the fibre.

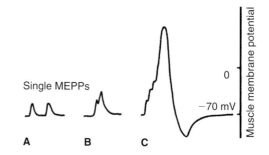

**Figure 3.2** Generation of an action potential at the neuromuscular junction. **A**, single miniature end-plate potentials (MEPPs). **B**, summation of two quanta but not sufficient to initiate an action potential. **C**, summation of sufficient MEPPs to initiate an action potential.

The drug *curare* acts as a competitive inhibitor, binding to the ACh receptors, preventing depolarization and causing paralysis of the muscle. The action of cholinesterase, hydrolysing ACh, is inhibited by *eserine* and *neostigmine*. These drugs potentiate neuromuscular junction transmission by prolonging the lifetime of ACh within the synaptic cleft and thus increasing the chance of its binding to an ACh receptor and depolarizing the muscle membrane.

Once hydrolysed, the free choline is transported back into the presynaptic axon terminal where it is resynthesized into ACh. This transport can be blocked with *hemicholium*, which produces a gradual paralysis as the ACh stored in the presynaptic terminal becomes exhausted with activity.

Anticholinesterases are used as an antidote to curare and some of the nerve gas poisons that also inactivate the ACh receptors. They are also used in the management of myasthenia, where antibodies against the ACh receptors block transmission (see Ch. 15). Curare is used in surgery as a muscle relaxant, as is *suxamethonium*. Unlike curare, suxamethonium causes the postsynaptic membrane to become depolarized so that the membrane passes into an inexcitable refractory state.

## PROPAGATION OF THE ACTION POTENTIAL

Action potentials initiated at the neuromuscular junction propagate along the length of the fibre and the T tubules, which are an extension of the surface membrane, into the interior of the muscle fibre. As the action potential spreads into the interior of the fibre, the T tubular membrane is depolarized at a time when the surface membrane is repolarizing, and this difference in potential could lead to the development of local circuits and repetitive firing (myotonia). However, muscle fibres have a high chloride conductance, which makes the muscle membrane more difficult to stimulate and thus reduces the chances of repetitive firing (Adrian & Marshall 1976). In some clinical conditions (e.g. myotonia congenita), the chloride conductance is reduced, leading to repetitive firing and slow relaxation of muscle (Adrian & Bryant 1974; see Ch. 15).

## EXCITATION–CONTRACTION COUPLING

As the wave of depolarization passes down the T tubules there is an interaction with the sarcoplasmic reticulum (SR) that results in the release of calcium (Fig. 3.3), initiating the interaction of actin and myosin and muscle contraction. This process is known as excitation–contraction coupling (EC coupling). The sequence of events leading from

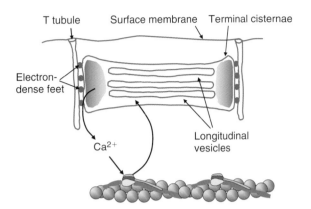

**Figure 3.3** Diagrammatic representation of the arrangement of sarcoplasmic reticulum and T tubules. Calcium is stored and released from the terminal cisternae, interacts with troponin C, and is then re-accumulated in the longitudinal vesicles of the sarcoplasmic reticulum.

excitation to contraction is of critical importance but the mechanisms involved have been, until recently, one of the least well understood aspects of muscle physiology.

### Structure of the triad

Where the T tubules meet the SR, the two membranes are opposed to one another and, because there is usually one portion of SR membrane on each side of the T tubule, the structure, when seen under the EM, is known as a *triad* (see T tubular system in Ch. 1 and Fig. 1.8). Close observation of the junction between the T tubular membrane and the SR, using high-power EM, shows that although the two membranes are separate there are darker regions, known as *electron-dense feet*, that span the gap. Even closer observation shows that the electron-dense feet consist of two components, one originating in each membrane. These structures have been identified by substances that bind to them. The structures in the T tubular membrane bind *dihydropyridines* (DHPs), which act as calcium channel blockers, and the structures are thus known as *DHP receptors*. The structures in the SR membrane bind a dye called *ryanodine* and are therefore known as *ryanodine receptors*. The DHP receptors are arranged in an orderly fashion on the T tubular membrane and each consists of four subunits. The ryanodine receptors are also

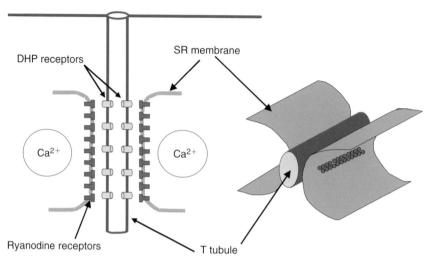

**Figure 3.4** Arrangement of dihydropyridine (DHP) and ryanodine receptors on the T tubular and sarcoplasmic reticulum (SR) membranes. Note that the T tubular membrane is continuous with the surface membrane of the muscle fibre, whereas the SR is an internal membrane system containing the calcium.

composed of four subunits in the SR membrane and are placed opposite DHP receptors, but there are only half as many DHP as ryanodine receptors and consequently only every other ryanodine receptor is matched to a DHP receptor (Peachey & Franzini-Armstrong 1983) (Fig. 3.4). The ryanodine receptors are very large and project into the space between the two membranes, making up the main bulk of the electron-dense feet.

## Calcium release from the sarcoplasmic reticulum

Calcium stored in the SR is released through the ryanodine receptors into the narrow gap between the SR and T tubular membranes, and then diffuses into interior of the fibre where it binds to troponin C on the thin filaments.

In skinned fibre preparations the SR can be loaded with calcium, which is retained against large concentration gradients. Paradoxically, this calcium is released in response to an increase in calcium concentration on the outside of the SR membrane. Thus a small amount of calcium entering the cell during an action potential can act as a trigger causing the release of calcium stored in the SR. In cardiac muscle a large portion of the inward current of the action potential is carried by

calcium, the function of which is to trigger calcium release from intracellular stores in the SR. Cardiac muscle is thus dependent on extracellular calcium to trigger the release of calcium required for contraction, and contractions fail if the extracellular calcium is removed. Although skeletal muscle SR has the same properties as cardiac SR, skeletal muscle can continue to contract for an appreciable time when external calcium is removed, so it is unlikely that calcium entry from the extracellular space plays a major role in the activation of skeletal muscle; some other way of releasing calcium from the SR is required.

In a number of tissues, calcium is released as a consequence of the hydrolysis of phosphatidylinositol, the inositol 1,4,5-triphosphate ($IP_3$) formed being able to release calcium from internal stores. $IP_3$ may have a role in smooth muscle, but has little effect in skinned skeletal muscle fibres (Posterino & Lamb 1998) and is therefore unlikely to be important in intact skeletal muscle.

The DHP receptors in skeletal muscle are specialized calcium channels. Normal L-type calcium channels have a voltage sensor that opens the channel when depolarized. In skeletal muscle the voltage sensor in the DHP receptors is linked, possibly mechanically, to the ryanodine receptor in the SR membrane. When the T tubular membrane

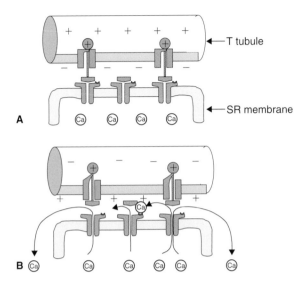

**Figure 3.5** Release of calcium from the sarcoplasmic reticulum (SR) in response to depolarization of the T tubular membrane. **A**, with a normal resting potential the DHP voltage sensor blocks calcium movement through the ryanodine receptor. **B**, when the T tubular membrane depolarizes, the voltage sensor moves outwards unblocking the channel, and calcium floods out. Ryanodine receptors may also be activated by calcium binding to a calcium receptor, initiating calcium-induced calcium release.

is depolarized, the movement of the DHP voltage sensor opens the ryanodine channel, allowing calcium to move from the interior of the SR through the narrow gap between the T tubular and SR membranes and then into the interior of the muscle fibre (Fig. 3.5) (Chandler et al 1976a,b, Dulhunty et al 1996, Melzer et al 1995).

The calcium released as a result of the interaction of DHP and ryanodine receptors may be the total required for activation of the myofilaments, but it is also possible that calcium released from one site on the SR may stimulate release of calcium from an adjacent site by calcium-induced calcium release, thus amplifying the release initiated by the voltage sensors of the DHP receptors.

## Factors affecting calcium release

The amount of calcium released in response to a single action potential may vary. Some substances potentiate the release of calcium and thus the size of the twitch (*caffeine, zinc, thiocyanate*), whereas others depress it (*dantrolene*). Caffeine

is particularly interesting because it has been used in experiments concerning the mechanism of fatigue. The ryanodine receptor has a caffeine binding site, and with low doses ($1-2\,\text{mmol}\,\text{L}^{-1}$) the caffeine potentiates the normal calcium release process but high concentrations (approximately $20\,\text{mmol}\,\text{L}^{-1}$) open the channel and calcium floods out causing the muscle to go into a contracture, a state where force is generated in the absence of any electrical activity. In certain circumstances *halothane*, a commonly used surgical anaesthetic, can have a similar effect. In normal subjects there is no noticeable effect, but in some susceptible patients the anaesthetic causes a large and dangerous release of calcium (see Ch. 15). Calcium release is also decreased with fatigue, and this is considered at some length in Chapter 10.

## Calcium transients

Intracellular calcium movements can be measured using a number of compounds that either emit light or fluoresce in the presence of calcium. The first to be used was the protein aequorin obtained from jellyfish and, subsequently, fluorescent compounds such as quin 2 and fura 2 have been developed from the calcium buffer EDTA. With skeletal muscle the indicators are microinjected into individual fibres, either single fibres that have been dissected from amphibian muscles or superficial fibres of small bundles of mammalian muscle. The stoichiometry and kinetics of calcium binding to the indicators are complex but, to a first approximation, the light signal gives an indication of the magnitude and time course of calcium release (Fig. 3.6).

## REGULATORY PROTEINS

The interaction of actin and myosin in mammalian skeletal muscle is regulated by tropomyosin and the troponin complex. The generally accepted mechanism for this regulation is the 'steric blocking' model. At rest, the tropomyosin is positioned by the troponin complex so that it covers the myosin binding sites on the actin monomers preventing the formation of cross-bridges. Calcium binding to troponin C causes a change in conformation, which moves the tropomyosin so that the actin binding sites are exposed (Fig. 3.7), cross-bridges are formed and force develops. The tropomyosin spans seven actin monomers, and

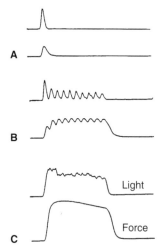

A

B

C

**Figure 3.6** Calcium release and force. Superficial fibres of a mouse diaphragm preparation were injected with aequorin. The light signal gives an indication of the intracellular free calcium concentration. **A**, single twitch; **B**, stimulated at 20 Hz; **C**, stimulated at 80 Hz. The upper trace is the light signal, the lower trace the force.

one troponin complex is responsible for controlling the activity of all seven subunits.

## REMOVAL OF CALCIUM FROM THE INTERIOR OF A MUSCLE FIBRE

Calcium is pumped back into the sarcoplasmic reticulum by a mechanism that requires ATP. The calcium ATPase of the sarcoplasmic reticulum is $Mg^{2+}$ dependent and transports two $Ca^{2+}$ into the SR in exchange for one $H^+$ and at the cost of one ATP molecule. The activity of individual enzymic sites is relatively low, but the high density of pumps in the SR membrane ensures that the cytosolic calcium can be rapidly reduced to very low levels, around $10^{-8}\,mol\,L^{-1}$, and in so doing removes calcium from troponin, leading to relaxation. Calcium accumulates in the central portions of the SR (longitudinal vesicles) but is stored in the terminal cisternae in association with the calcium binding protein *calsequestrin*. The internal calcium concentration is around $1\,mmol\,L^{-1}$ in the terminal cisternae.

Actively respiring mitochondria will accumulate calcium ions, but it is not known whether

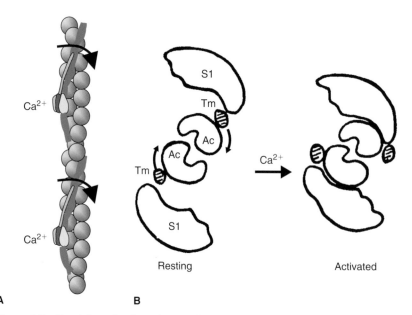

**Figure 3.7** Regulation of actin and myosin interaction. **A**, arrows show the movement of tropomyosin (Tm) when calcium is bound to troponin. **B**, cross-section of the filament showing how the movement of tropomyosin allows binding of the myosin head (S1) to actin (Ac).

this accumulation plays a significant role in the relaxation of normal muscle.

In muscles of amphibians and small mammals there is a high concentration of the protein *pavalbumin*. This is another calcium binding protein and may have a role in buffering the internal calcium concentration or speeding relaxation, acting in parallel with the SR to remove calcium from troponin C. However, pavalbumin is present in very low concentrations in human muscle and probably plays little or no part in determining the speed of relaxation.

## TWITCHES AND THE SUMMATION OF FORCE TO GENERATE A TETANUS

A single action potential will release a quantity of calcium from the SR, which binds to troponin C leading to the generation of force. At the same time the free calcium is being pumped back into the SR, so the calcium concentration remains raised for a short time and cross-bridges that detach find there are no free actin binding sites at which to reattach. This process generates the characteristic 'twitch', the shape and size of which depend on a number of factors. The rate of rise of the twitch will depend partly on the amount and speed of calcium release and the facility with which it binds to troponin. The speed with which the myosin heads bind with actin and generate force is another factor, as is the compliance of the system. The more series elasticity, the longer it will take to stretch these elements until an appreciable force is generated. The relaxation phase of the twitch is likewise affected by a variety of factors. The first is the rate at which calcium is pumped back into the SR, and a second is the rate at which cross-bridges dissociate. The compliance of the series elements will, again, affect the rate of force loss. The timing and height of the twitch therefore depends on a number of factors including the kinetics of the cross-bridges as well as those of both calcium release and reuptake.

The force records in Figure 3.8 show that, when stimulating this fast mouse muscle at 20 Hz, the force almost returned to baseline before the next stimulus, resulting in a series of twitches. When the interval between twitches was reduced to

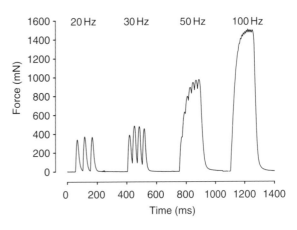

**Figure 3.8** Twitch summation and the development of titanic force (mouse medial gastrocnemius, 37°C). Stimulated at 20, 30, 50 and 100 Hz.

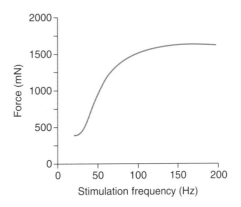

**Figure 3.9** Force–frequency relationship for the fast mouse muscle data illustrated in Figure 3.8. Note that peak force has been used for the unfused contractions.

33 ms (30-Hz stimulation), force was still increased when the next stimulus arrived and the twitches summated, generating an unfused tetanus. As the frequency increased further, the degree of summation increased giving a smoother tetanus until, at 100 Hz, with only 10 ms between stimuli, the tetanus showed only a small ripple on the plateau of force. Stimulating at even higher frequencies produces only slightly higher forces as the interior of the muscle fibre becomes saturated with $Ca^{2+}$, fully activating the myofibrils. As the frequency increases, so the oscillation on top of the record decreases and the relationship between the degree of oscillation and the mean force at a given

frequency is sometimes used as an indicator of the speed of the muscle (e.g. Gerrits et al 1999). The conventional way of presenting such data is to plot the force–frequency relationship (Fig. 3.9). A decision needs to be made whether to use the mean force or the peak force. For high frequencies this makes little difference, but at low frequencies the mean and peak forces can be very different. Figure 3.9 shows the relationship between peak force and frequency.

## REFERENCES AND FURTHER READING

Adrian R H, Bryant S H 1974 On the repetitive discharge in myotonic muscle fibres. Journal of Physiology 240: 505–515

Adrian R H, Marshall M W 1976 Action potentials reconstructed in normal and myotonic muscle fibres. Journal of Physiology 258: 125–143

Blinks J R, Rudel R, Taylor S R 1978 Calcium transients in isolated amphibian skeletal muscle fibres: detection with aequorin. Journal of Physiology 277: 291–323

Chandler W K, Rakowski R F, Schneider M F 1976a Non linear voltage dependent charge movement in frog skeletal muscle. Journal of Physiology 254: 245–283

Chandler W K, Rakowski R F, Schneider M F 1976b Effects of glycerol treatment and maintained depolarization on charge movement in skeletal muscle. Journal of Physiology 254: 285–316

Dulhunty A F, Junankar P R, Eager K R, Ahern G P, Laver D R 1996 Ion channels in the sarcoplasmic reticulum of striated muscle. Acta Physiologica Scandinavica 156: 375–385

Endo M 1977 Calcium release from the sarcoplasmic reticulum. Physiological Reviews 57: 71–108

Gerrits H L, de Haan A, Hopman M T E, van der Woude L H V, Jones D A, Sargeant A J 1999 Contractile properties of the quadriceps muscle in individuals with spinal cord injury. Muscle and Nerve 22: 1249–1256

Gonzalez-Serratos H 1983 Inward spread of activation in twitch skeletal muscle fibres. In: Peachey L D, Adrian R H, Geiger S R (eds) Handbook of physiology, section 10: skeletal muscle, pp 325–353. Bethesda: American Physiological Society

Hashimoto T, Hirata M, Ito T, Kanmura Y, Kuriyama H 1986 Inositol 1,4,5-triphosphate activates pharmacomechanical coupling in smooth muscle of the rabbit mesenteric artery. Journal of Physiology 370: 605–618

Heizmann C W, Berchtold M W, Rowlerson A M 1982 Correlation of parvalbumin concentration with relaxation speed in mammalian muscles. Proceedings of the National Academy of Sciences 79: 7243–7347

Melzer W, Herrmann-Frank A, Lüttgau H C 1995 The role of $Ca^{2+}$ ions in excitation–contraction coupling of skeletal muscle fibres. Biochimica et Biophysica Acta 1241: 59–116

Peachey L D, Franzini-Armstrong C 1983 Structure and function of membrane systems of skeletal muscle cells. In: Peachey L D, Adrian R H, Geiger S R (eds) Handbook of physiology, section 10: skeletal muscle, pp 23–71. Bethesda: American Physiological Society

Posterino G S, Lamb G D 1998 Investigation of the effect of inositol triphosphate in skinned skeletal muscle fibres with functional excitation–contraction coupling. Journal of Muscle Research and Cell Motility 19: 67–74

## QUESTIONS TO THINK ABOUT

1. What is the consequence of increased extracellular potassium or decreased sodium concentration for (a) the resting membrane potential and (b) the action potential?

2. Summarize the events from excitation of the T tubule by an action potential to the release of calcium by the sarcoplasmic reticulum.

3. What happens to the calcium released per stimulation pulse when the stimulation frequency is increased?

4. Which processes lead to actin–myosin interaction following the release of calcium from the sarcoplasmic reticulum?

5. How is calcium removed from the free pool in the sarcoplasm?

6. What determines whether or not fusion of force occurs as a result of consecutive stimulus pulses?

7. Why does force reach steady values at high stimulation frequencies?

8. Caffeine potentiates the twitch response. Argue whether a similar force enhancement could be expected for contractions at higher stimulation frequencies.

9. What would be the consequence of a faster removal of calcium from the cytoplasm for (a) relaxation and (b) the stimulation frequency–force relation?

10. Calcium is pumped back into the sarcoplasmic reticulum right from the start of a contraction, and this process costs a lot of energy. Why is the re-accumulation not delayed until the end of a contraction?

# 4

# Fibre types and motor units

As early as the seventeenth century a number of authors had commented on the fact that muscles differ in their appearance, but it was not until 1873 that Ranvier, the French physician and physiologist, recognized that skeletal muscles not only differ in colour but also have different contractile properties. We recognize this difference at the table when we talk about white and dark meats. Chicken and turkey breast is 'white' meat, whereas the legs have 'dark' meat on them. It is sometimes surprising that other birds, pigeon for instance, have dark breast muscles, but we will see that this is a consequence of the different use of their flight muscles. Chickens and turkeys are virtually flightless, using their breast muscles for the occasional burst of activity to get on to a perch, whereas the pigeon is an endurance athlete capable of travelling great distances. There are also differences between domesticated and wild animals. Domesticated pork is a pale meat, whilst wild boar is dark red, as is the meat of most 'game' that has to survive the dangers of the chase. The red colour of meat is due to the presence of iron in myoglobin and in mitochondrial cytochrome oxidase. Consequently the dark red colour indicates a muscle high in oxidative potential.

Although some animal muscles have a uniform appearance, most—including human muscles—have a mixed composition, being made up of fibres with differing contractile properties and staining reactions. It will become apparent that these differences are due to the innervation of the fibres and the different ways in which they are used. We will first discuss the way in which nerves innervate muscle.

## THE MOTOR UNIT

Each mature mammalian muscle fibre is inner-vated by a single axon; one α motoneuron will supply through its axonal branches many fibres scattered throughout the muscle. In a healthy muscle the innervation is almost entirely random and adjacent muscle fibres are most likely to be supplied by branches from different motoneu-rons (Fig. 4.1). All the muscle fibres supplied by one motoneuron form a *motor unit* and the por-tion of the muscle containing the motor unit is known and the territory of that motoneuron.

A single impulse originating in a motoneuron passes down the axonal branches and stimulates all the muscle fibres at virtually the same moment, generating a synchronous action potential in all the fibres and a twitch of force. The size of the action potential recorded from surface electrodes is the sum of the individual potentials and depends on the number of fibres in the motor unit. Likewise, the size and speed of the twitch depends on the number of muscle fibres in the unit and their con-tractile properties. Within a single muscle there is

a range of motor unit size, the number and type of muscle fibres within each being determined largely by the size and function of the motoneuron that innervates them.

The number of motor units in a muscle and the number of fibres within each motor unit var-ies from muscle to muscle. The tendency is for muscles required for fine motor control to have smaller motor units. For example, the small eye muscles have large numbers of very small motor units, with about 10 fibres per unit. In the first dor-sal interosseous of the hand there are estimated to be around 120 motor units with about 300 fibres per unit, while in the large weight-bearing medial gastrocnemius there are around 600 motor units and approximately 2000 fibres per unit.

## HISTOCHEMISTRY

The differences in appearance between white and dark muscles seen at the gross anatomical level are also evident at the level of the individual fibres, but it is difficult to detect much structure

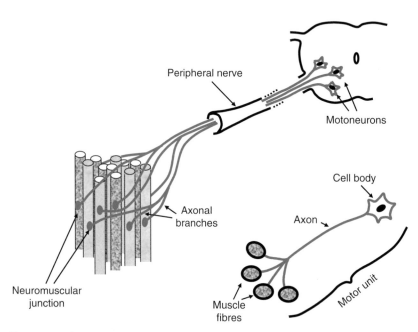

**Figure 4.1** Concept of the motor unit. A motor unit consists of the motoneuron and all the scattered muscle fibres that it innervates.

in sections of unstained muscle using conventional light microscopy. For this reason a variety of stains is commonly used on thin frozen sections to demonstrate differences between muscle fibres. Histochemical staining reveals different enzymic activities, such as myosin ATPase or mitochondrial enzymes such as succinate dehydrogenase, or can be used to identify substances such as glycogen or lipid. More recent advances have used immunological techniques to identify specific proteins by their reaction with labelled antibodies and, even more recently, individual mRNAs can be localized using in situ hybridization techniques.

## Myosin ATPase activity

One of the first studies that explored the biochemical basis for differences in function in skeletal muscle was that of Bárány (1967), essentially repeating the observations of Ranvier but using modern quantitative methods. He measured maximally activated myosin ATPase activity at physiological pH in a wide range of muscles and found that the ATPase activity varied severalfold and was directly proportional to the speed of shortening of the muscle (Fig. 4.2). The faster the rate of ATP turnover and energy expenditure, the quicker the muscle could shorten.

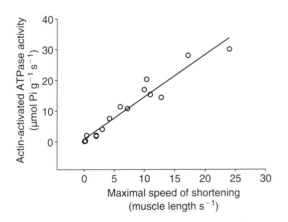

**Figure 4.2** Relationship between maximal shortening velocity and actin-activated myosin ATPase activity from various animal species. (Redrawn from the data of Bárány 1967.)

## Myosin ATPase: pH stability

One of the most common stains used to differentiate between fibre types makes use of the different stabilities of the various myosins when incubated in acid or alkali media.

The procedure needs to be described in two parts. The first stage consists of incubating the section at either an alkaline or an acid pH. Although the values vary somewhat from species to species, in general, fast myosin is inactivated at acid pH (about pH 4.3–4.6) whereas slow myosin is inactivated at alkaline pH (around pH 9.4). Having preincubated the section and inactivated the ATPase activity in one set of fibres, the remaining activity can be visualized by incubating the section with ATP and an excess of calcium at a slightly alkaline pH so that the inorganic phosphate liberated precipitates as calcium phosphate. By a further series of manipulations the calcium phosphate can be visualized, usually as a dark brown or black colour. The myosin ATPase revealed in this way is often referred to as being of either 'high' or 'low' activity. It is important to remember, however, that the ATPase staining tells us *nothing* about the *activity* of myosin ATPase at physiological pH; it only gives information about the *stability* of the enzyme at the pH used for the first inactivation step. The faster myosins tend to be stable at alkaline pH and labile in acid, while for the slower myosins it is the other way round. Two major groups of fibre are identified in human muscle, type 1 and type 2. Type 2 fibres are further subdivided into what are conventionally called type 2a and 2b (although see below). A fourth fibre subgroup (type 2c) represents myosin found largely in the developing embryo and in regenerating fibres.

For mouse and rat muscles, use of this type of staining approach presents a more complex picture, with at least four types being seen; the additional type is known as 2x. Examples are shown of myosin ATPase staining for rat muscle in Figure 4.3A,B. After acid preincubation (pH 4.6) (Fig. 4.3A), myosin ATPase activity is best maintained in the slowest type 1 fibres (dark staining), whereas the type 2a fibres (light staining) lose all their myosin ATPase activity. Intermediate staining is obtained for both the type 2x and 2b fibres.

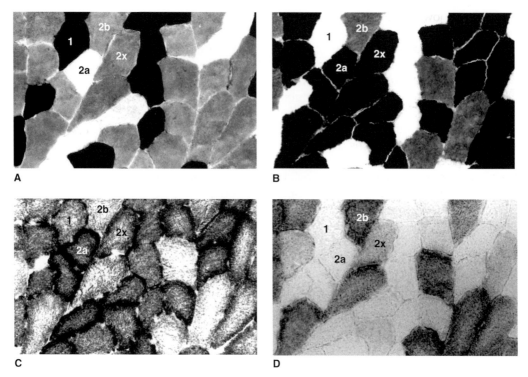

**Figure 4.3** Serial sections of rat medial gastrocnemius muscle. **A & B**, stained for myosin ATPase activity after acid and alkaline preincubation, respectively. **C**, stained for succinate dehydrogenase activity. **D**, stained for α-glycerol phosphate dehydrogenase activity.

In contrast, after alkaline preincubation (pH 9.4) (Fig. 4.3B) the type 1 fibres (light staining) lose all their activity. The type 2a fibres stain dark, as do the type 2x fibres, having retained most of their myosin ATPase activity. The type 2b fibres show intermediate staining, as after acid preincubation. Note that only by combining both stains is it possible to identify the type 2x fibres (Table 4.1).

Immunological studies and staining with specific antibodies to the different types of heavy myosin have shown that the human type 2b is, in fact, not the same as the rodent 2b myosin, but rather is homologous with the rat 2x myosin. It seems that human muscle lacks the very fastest type 2b myosin. Strictly speaking, human fibres that have been generally referred to as type 2b should be called 2x; this nomenclature is adopted in the more recent literature.

**Table 4.1** Rodent muscle fibre types stained for myosin ATPase after preincubation at pH 4.6 or 9.4

| | Fibre type | | | |
|---|---|---|---|---|
| Preincubation | 1 | 2a | 2x | 2b |
| pH 4.6 | + + | 0 | + | + |
| pH 9.4 | 0 | + + | + + | + |

+ +, Strong positive stain; +, intermediate stain; 0, unstained.

## Myosin heavy chains

The conventional myosin ATPase staining method frequently shows fluctuations in staining between fibres of apparently the same 'type', especially type 2a, and these differences are sometimes ascribed to variations in the temperamental staining process.

However, there is now another explanation, as immunological identification of different myosins has shown that rather than fibres expressing only one myosin type, as was previously thought, there can be co-expression of at least two myosin types. Hybrid fibres containing different myofibrillar protein isoforms are quite frequent and may represent the rule rather than the exception (Staron & Pette 1993). Multiple myosin heavy chains (MHCs) are regularly co-expressed during development, during transformation induced by electrical stimulation, and during changes in hormonal state and physical training. In rat muscle, fibres are seen that contain both 2a and 2x, and 2x and 2b, in all proportions. In human muscle, where the true 2b myosin seems to be absent, approximately 3% of fibres have been found to express both type 1 and 2a, and up to 40% of fibres may express 2a and 2x myosin (Larsson & Moss 1993).

## Myosin light chains

Associated with every pair of myosin heavy chains there are two pairs of myosin light chains (MLCs); these exist in a variety of forms and combinations in different fibre types. One of the pairs consists of the 'essential' or 'alkali' light chains, the latter name given because they are soluble in alkaline solution. The other pair are the regulatory light chains, so called because their function in smooth muscle is to regulate the interaction of myosin with actin depending on their state of phosphorylation. Skeletal muscle has a different control system, but the regulatory MLCs probably retain some function, possibly in relation to post-tetanic potentiation. There are two types of fast essential light chain, MLC-1f and MLC-3f, and two slow forms, MLC-1sa and MLC-1sb. MLC-1sa is the predominant essential light chain found in human skeletal muscle. The regulatory light chains exist in two forms, the fast MLC-2f and the slow MLC-2s.

## Actin

There are two forms of actin, α-skeletal actin and α-cardiac actin and, whilst both are expressed in adult skeletal muscle, there is no suggestion that there are any differences between different fibre types (Schiaffino & Reggiani 1996).

## Tropomyosin

Tropomyosin (Tm) exists as a dimer of α and β subunits, and can form both homo- and heterodimers. In addition, the Tm-α subunit is found in fast and slow forms. Fast fibres tend to have a predominance of the Tm-β subunits together with Tm-α-fast (Schiaffino & Reggiani 1996).

## Troponin

There are three troponin (Tn) proteins: TnC, the calcium binding protein; TnI, the inhibitory subunit; and TnT, the subunit that binds to tropomyosin. All exist in at least two forms, fast and slow, found in different skeletal muscle fibre types (Schiaffino & Reggiani 1996).

## Mitochondrial enzyme activities

Stains for mitochondrial activity require a substrate that is oxidized and an electron acceptor that changes colour when reduced. A commonly used combination is NADH as substrate and tetrazolium blue as the electron acceptor, giving a blue precipitate. The flavoprotein enzyme system (complex I; see Fig. 6.5) that reduces tetrazolium blue is termed NADH tetrazolium reductase (NADH-TR) (Fig. 4.4). The use of succinate as substrate with tetrazolium blue (Fig. 4.3C) gives a measure of complex II activity (succinate dehydrogenase, SDH; see Fig. 6.5).

Fibres with a high mitochondrial content also store intracellular lipid, which can be seen as fine droplets scattered throughout the fibre between the myofibrils. This represents a considerable store of fuel in the form of triglyceride.

## Glycolytic activity

It is possible to visualize a number of enzyme activities as indicators of glycolytic potential: lactate dehydrogenase, phosphofructokinase, pyruvate kinase. One commonly measured enzyme is myophosphorylase (Fig. 4.4), the first enzyme in the glycolytic pathway, which breaks the complex polymer glycogen down to simple units of glucose 1-phosphate. Another enzyme used to distinguish fibres based on their relative glycolytic potential is α-glycerol phosphate dehydrogenase (α-GPDH) (Fig. 4.3D). Although this enzyme is not part of the direct chain of glycolytic reactions, it is related to the glycolytic pathway because of its action shuttling reduction equivalents into the mitochondria where they are used for energy production and regenerating NAD (see Fig. 6.3).

Glycogen is stored within muscle fibres as small granules visible on electron microscopy, but it can also be visualized with an iodine-based stain. There is a wide variation between muscle fibres in their content, but generally the more glycolytic fibres have the higher content.

## Relationship between different histochemical stains

When muscle sections are stained for a range of enzymic activities, they tend to fall into groups. Figure 4.4 shows serial sections of human muscle. In muscle that has not undergone specific oxidative training, type 1 fibres (as defined by their myosin ATPase) have the greatest oxidative capacity and lowest glycolytic activity, and type 2b

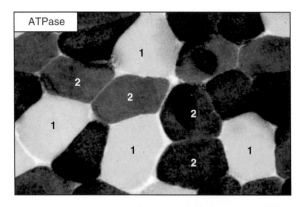

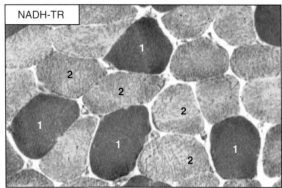

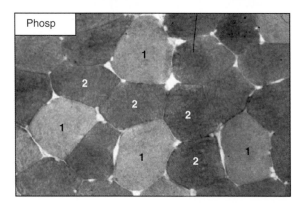

**Figure 4.4** Serial sections of human quadriceps muscle. Sections stained for myosin ATPase (preincubation pH 9.4), NADH tetrazolium reductase (NADH-TR), and phosphorylase. Type 1 and 2 fibres that can be seen in each section are labelled.

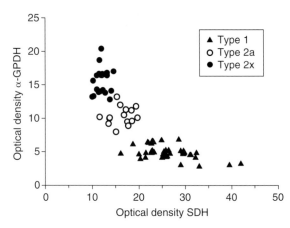

**Figure 4.5** Relationship between succinate dehydrogenase (SDH) and α-glycerol phosphate dehydrogenase (α-GPDH) activities. Data are from analysis of a biopsy from human quadriceps muscle. Optical density indicates the intensity of staining; fibre types of the individual fibres are shown, based on myosin ATPase activity.

(type 2x in human muscle) fibres are the opposite, with low mitochondrial content and high glycolytic enzymes (see also Fig. 4.3 for rat muscle). The type 2a and 2x fibres, together with the various mixtures of 2a and 2x, are intermediate in their content of glycolytic and oxidative enzymes. This reciprocal relationship can be seen for human muscle in the data presented in Figure 4.5.

It is notable that, although fibres can be broadly grouped into different types, there is overlap between the groups, especially with respect to oxidative activity.

## CONTRACTILE PROPERTIES

Not only are there correlations between the expression of different enzyme activities in muscle fibres, but there are parallels with function and contractile characteristics. These were alluded to in the introduction to this chapter where differences were noted in the appearance and function of chicken and pigeon breast muscles.

White muscles develop force rapidly and also relax more rapidly than do the darker red muscles. This can be illustrated for two muscles of the lower leg: the red soleus that forms part of the calf muscle group, and the white extensor digitorum longus, found at the front of the leg and which has the task of raising the toes. When stimulated at a frequency of 10–20 Hz, the fast extensor digitorum longus muscle reacts quickly enough for the force to fall back to baseline before the next impulse. However, for the slower soleus muscle, the next impulse comes before relaxation is complete and the contraction is superimposed on the force remaining from the previous stimulus. In this way the individual twitches are said to summate or fuse. When stimulated at a sufficiently high frequency, the muscle produces a smooth plateau of force. The frequency required to achieve this plateau is known as the fusion frequency, and is higher for fast as compared to slow muscles (Fig. 4.6).

In mammalian muscles the twitch force is usually about one-fifth to one-tenth of the maximum tetanic force, and in this respect it differs from amphibian muscle where the twitch force is around 80% of the maximum force. The relatively small twitch of mammalian muscle is probably due to a smaller quantity of calcium being released per action potential and may allow a finer grading of contraction force.

There are a number of features that distinguish the contractile properties of fast and slow muscles:

- *The shape of the twitch*: As shown in Figure 4.6, the twitch of a fast muscle has an earlier peak and more rapid relaxation than that of a slow muscle. Measurements are commonly made of contraction time and half relaxation time.
- *Relaxation from an isometric tetanus*: Figure 4.6 shows the relaxation phase from a tetanus. The last half of the curve approximates a single exponential, and a useful measure of speed is the half time of this portion. The faster muscle has a shorter half relaxation time.
- *Fusion frequency*: The frequency at which a smooth contraction is generated is higher in fast muscles (Fig. 4.6), and consequently the force–frequency relationships are quite distinctive (Fig. 4.7).

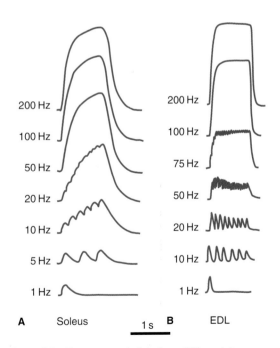

**Figure 4.6** Force generated during a 500-ms tetanus at different frequencies (isolated mouse muscles, 25°C). **A**, soleus; **B**, extensor digitorum longus (EDL).

- *Fatiguability*: Muscles vary widely in their response to prolonged activity. Fast muscles fatigue much more rapidly than slower muscles.

As first noted by Ranvier, there is a clear connection between the appearance and contractile properties of certain skeletal muscles, such as the red and white muscles of a chicken or the soleus and extensor digitorum longus in a mouse. These muscles are, however, somewhat unusual in consisting predominantly of one fibre type, whereas the majority of skeletal muscles are a mixture of different types. It is important to know whether the different fibre types within a single muscle also have different contractile properties.

## RELATIONSHIP BETWEEN HISTOCHEMISTRY AND CONTRACTILE PROPERTIES OF MOTOR UNITS

A technique that has been used to investigate this relationship in single motor units in rat and cat muscles is illustrated in Figure 4.8 (Burke et al 1971, Kugelberg 1973). By stimulating a single axon or a single motoneuron, all the fibres in one motor unit can be made to contract simultaneously, and by using sensitive force recording techniques the contractile characteristics (speed, fusion frequency, fatiguability) of this motor unit can be determined. The maximum tetanic force generated gives an indication of the number of fibres comprising the motor unit. On the basis of

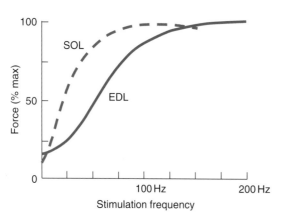

**Figure 4.7** Relationship between force and stimulation frequency. Isolated mouse soleus (SOL, slow) and extensor digitorum longus (EDL, fast), 25°C.

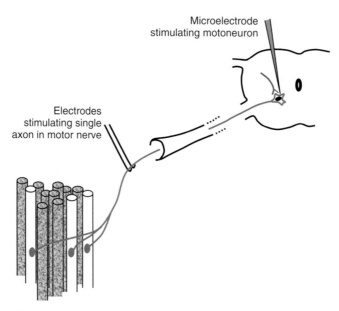

**Figure 4.8** Identification of fibres belonging to a single motor unit. Repetitive stimulation of the motoneuron or axon depletes the fibres belonging to that motor unit of glycogen.

size, speed and fatiguability, motor units are found to fall between two extremes: large, fast and fatiguable, or small, slow and fatigue resistant (Fig. 4.9). The strength of the fast motor units, is greater than that of the slower units, primarily because the fast motor units contain more fibres.

Having measured the size and contractile characteristics of the motor unit it is necessary to determine the histochemical properties of the muscle fibres of which it is composed. What is required is some way of identifying which fibres within the muscle have been contracting; this is done by stimulating the unit repetitively to deplete the fibres of glycogen. The muscle is

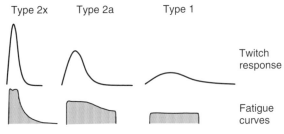

Type 2x     Type 2a     Type 1

Twitch response

Fatigue curves

**Figure 4.9** Contractile characteristics of the main types of motor unit. Upper traces are single twitches; lower traces show the fatigue curves during repetitive titanic stimulation. Note the differences in force, speed and fatiguability.

then removed, frozen, sectioned and stained for glycogen; the depleted fibres of the motor unit stand out as pale-staining cells against the dark background of glycogen-rich fibres (Fig. 4.10A). Serial sections stained for myosin ATPase, mitochondrial and glycolytic enzymes will then define the histochemical properties of the depleted fibres (Fig. 4.10B). In this way it is possible to relate the histochemical and contractile properties of the same fibres.

Examining their histochemical properties in animals, the larger and faster motor units tend to contain type 2b fibres whereas the small slow motor units are composed predominantly of type 1 fibres. Type 2a and 2x motor units span a range of size and fatigue resistance, which is reflected in the broad spectrum of their mitochondrial enzyme activities. The relationships between size, speed and fatiguability of the different types of motor unit are summarized in Figure 4.11.

Some of the older, and especially American, literature commonly used nomenclatures such as FG (fast, glycolytic), FOG (fast, oxidative, glycolytic) or SO (slow, oxidative), which correspond to 2b, 2a/2x and type 1, or may refer to FF (fast, fatiguable), FR (fast, fatigue resistant) or S (slow). These classifications combine information about contractile and histochemical properties, and are useful verbal descriptors, but it should be realized

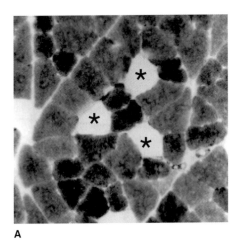

**A**

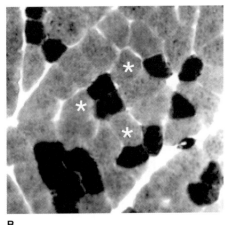

**B**

**Figure 4.10** Serial sections of rat muscle after a single motor unit was repetitively stimulated. **A**, stained for glycogen showing depletion in three fibres in this section. **B**, stained for myosin ATPase after preincubation at pH 9.4 (type 2x, dark; type 2b, grey) showing depleted fibres (*) to be type 2b (see Table 4.1).

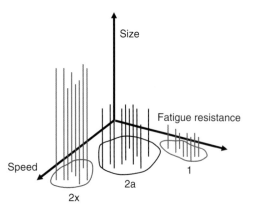

**Figure 4.11** Contractile properties of motor units. The size, speed and fatigue resistance of different types of motor unit, as defined by histochemical properties.

**Table 4.2** Summary of enzymic and contractile properties of major motor units

|  | Motor unit type | | |
| --- | --- | --- | --- |
|  | 1 | 2a | 2x |
| Myosin ATPase, pH 4.6* | + + | 0 | + |
| Myosin ATPase, pH 9.4* | 0 | + + | + + |
| Myosin ATPase, pH 7.0** | + | + + | + + + |
| Mitochondrial enzymes | + + + | + + | + |
| Glycolytic enzymes | + | + + | + + + |
| No. of fibres in motor unit | + | + + | + + + |
| Size of motoneuron | + | + + | + + + |
| Contractile speed | + | + + | + + + |
| Fatigue resistance | + + + | + + | + |

* Myosin activity after preincubation at the pH indicated.
** Activity of actin-stimulated myosin ATPase measured at physiological pH.

that most classifications oversimplify the situation. Although mammalian skeletal muscles all show similar varieties of fibres when classified by their histochemical properties, there are considerable differences between species with regard to contractile characteristics. Thus, a slow mouse muscle is still faster than the fastest human skeletal muscle. To some degree this is also true of fibres from different muscles in the same animal, type 1 fibres in the soleus muscle being slower than type 1 fibres in other skeletal muscles.

The properties of the main types of motor unit are summarized in Table 4.2.

Much of the early fundamental work on muscle physiology was based on amphibian muscles, and these also have a variety of fibre types; unfortunately the system of classification is the opposite of that used with mammalian muscle. Fast frog muscle fibres are type 1, while the slow variety are type 2.

## Contractile characteristics of human muscle fibres

The technique of inserting microelectrodes into motoneurons or of dissecting single axons from motor nerves is not well tolerated by human subjects, and consequently the relationship between histochemistry and contractile properties is less well established for human muscle fibres. Stephens

and co-workers (Garnett et al 1979) stimulated small intramuscular branches of a motor nerve, which they believed to be innervating single motor units in human muscle. After characterizing the contractile properties and depleting the unit of glycogen by repetitive stimulation, they biopsied the region of the muscle where the unit was thought to be located and used histochemical stains to characterize the fibres. Their results indicate that human fibre types have the same general relationship between contractile properties and histochemical staining as found in other mammalian species.

## MUSCLE FIBRE STRUCTURE

There are clearly major differences in the enzyme contents of different fibre types, as can be demonstrated with various staining techniques and observations with light microscopy. However, the differences between fibres extend further and are visible in the very structure of the contractile elements and the associated structural and supportive elements.

### T tubular system

The T tubules run from the surface membrane and branch around and make contact with the

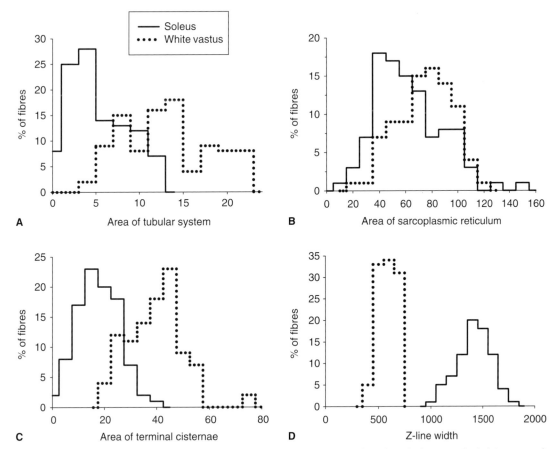

**Figure 4.12** (**A–D**) Histograms of surface areas of the T tubular system, sarcoplasmic reticulum, terminal cisternae and Z-line width of muscles from guinea-pigs; slow soleus muscle and fast white vastus lateralis. (Redrawn from the data of Eisenberg 1983.)

myofibrils. In fast fibres the branching is more extensive than in slow fibres, so that in the former case tubules may completely encircle a myofibril whilst in the latter there may be only a glancing contact. Fast fibres have a T tubular system that is about twice as extensive as that of slow fibres (Fig. 4.12A).

## Sarcoplasmic reticulum

The sarcoplasmic reticulum (SR) is closely associated with the T tubules, and consequently the terminal cisternae are proportionately well developed in fast fibres and the longitudinal portions of the SR also tend to be more developed in the faster fibres (Fig. 4.12B,C). The larger area

of SR in the faster fibres means that there are more calcium ATPase sites to transport calcium; there may also be a greater density of calcium transporters per unit area of SR membrane.

## Z-line structure

The Z line has a complex structure and varies in thickness, depending on the fibre type. Faster fibres have thinner Z lines (Fig. 4.12D).

## M lines

M lines are lines seen on electron microscopy to run across the A band. They are structural proteins that help keep the array of thick myosin filaments in the correct orientation. There are suggestions that slow fibres have five of these lines, compared with three in faster fibres.

## MOTONEURONS

The diversity of muscle fibre types is also reflected in the neurons innervating the motor units. Motoneurons supplying the larger, faster motor units have larger cell bodies, larger-diameter axons and a greater number of axonal branches innervating the larger number of individual muscle fibres. The size of the cell body is probably the reason why the fast motoneuron can support the greater number of fibres, having the greater capacity for protein synthesis needed to support the extensive branching axonal network. What is also interesting, however, is that the electrical properties of the motoneurons also differ. The large cell bodies of the fast motoneurons have relatively sparse afferent innervation and consequently receive relatively little excitatory input from spindles. This means that they are less likely to fire in response to descending inputs, which is in keeping with the size principle of motor unit recruitment. Although slow to be recruited, when they do fire, the larger neurons fire at higher frequencies than the smaller neurons supplying the slower motor units. Neuromuscular junctions also differ between fibre types. The motor end-plate regions of fast fibres are more extensive and complex in structure than those of slow fibres. It is not known whether this simply represents differences between what can be maintained by a large neuron compared to a smaller cell body, or whether it has some functional significance.

## INTEGRATION OF STRUCTURE AND FUNCTION

The majority of differences between muscle fibres described here have obvious implications for function. Thus, differences in myosin activity suit the fibre to rapid or slow movements, and there are differences in oxidative and glycolytic pathways that meet the demands of the different types of activity. There are, however, other adaptations that do not have such obvious applications. Why, for instance, do slow fibres have wider Z lines than fast fibres, or more M lines or smaller motor end-plates?

REFERENCES AND FURTHER READING

Bárány M 1967 ATPase activity of myosin correlated with speed of shortening. Journal of General Physiology 50: 197–218

Bottinelli R, Betto R, Schiaffino S, Reggiani C 1994 Unloaded shortening velocity and heavy myosin chain and alkali light chain isoform composition in rat single skinned muscle fibres. Journal of Physiology 478: 341–349

Burke R E, Edgerton V R 1973 Motor unit properties and selective involvement in movement. Exercise and Sport Science Reviews 3: 31–81

Burke R E, Levine D M, Zajac F E, Tsairis P, Engel W K 1971 Mammalian motor units: physiological–histochemical correlation of three types in cat gastrocnemius. Science 174: 709–712

Eisenberg B R 1983 Quantive ultrastructure of mammalian skeletal muscle. In: Peachey L D, Adrian R H, Geiger S R (eds) Handbook of physiology, section 10: skeletal muscle, pp 73–112. Bethesda: American Physiological Society

Garnett R A F, O'Donovan M J, Stephens J A, Taylor A 1979 Motor unit organization of human medial gastrocnemius. Journal of Physiology 287: 33–43

Kugelberg E 1973 Histochemical composition, contraction speed and fatiguability of rat soleus motor units. Journal of the Neurological Sciences 20: 177–198

Larsson L, Moss R L 1993 Maximum velocity of shortening in relation to myosin isoform composition in single fibres from human skeletal muscles. Journal of Physiology 472: 595–614

Ranvier M L 1873 Propriétés et structures différentes des muscles rouges et des muscles blancs, chez les Lapins et chez les Raies. Comptes Rendus des Académie de Sciences 77: 1030–1034

Schiaffino S, Reggiani C 1996 Molecular diversity of myofibrillar proteins: gene regulation and functional significance. Physiological Reviews 76: 371–423

Staron R S, Pette D 1993 The continuum of pure and hybrid myosin heavy chain-based fibre types in rat skeletal muscle. Histochemistry 100: 149–153

1. Make a list of the properties that could be used to classify different muscle fibre types. Which of these properties form a continuum and which can be used as discrete markers of fibre type?

2. Explain why a type 1 fibre stains dark with the ATPase stain after preincubation at low pH, whereas ATP turnover in these fibres during activity is lower than in the type 2 fibres.

3. Why might the cross-sectional area of type 1 fibres be smaller than that of type 2b fibres in rodents?

4. Explain why succinate dehydrogenase (SDH) staining looks granular while the ATPase reaction gives a more even stain.

5. Why is the twitch force of an S motor unit smaller than that of an FF motor unit?

6. Draw in one figure the force–frequency relationship for a fast and a slow motor unit.

7. Explain why fast type 2b fibres have a high anaerobic and low aerobic capacity while, in contrast, the type 1 fibres have a high aerobic and low anaerobic capacity.

8. Explain what functional relevance there is in the fact that the area of the sarcoplasmic reticulum and the T tubular system is larger in type 2 compared with type 1 fibres.

9. Explain the functional significance of the relationship observed between SDH and $\alpha$-GPDH activity in muscle fibres.

10. What is the evidence that muscle fibres belonging to a single motor unit have the same histochemical properties? What are the implications of this evidence when thinking about the factors that determine fibre-type characteristics?

# 5

# Recruitment and control of motor units

We have seen in the preceding chapter that there is a range of motor units with distinct contractile and metabolic properties and it has been implied that these properties are consistent with specialized roles such as in frequent, slow, low-force postural movements or occasional, rapid, high-power activity such as jumping and sprinting. This chapter reviews the evidence that this is the case.

## MOTOR UNIT RECRUITMENT

It is not an easy task to identify motor units within a muscle that are, or have been, active during a movement, but there are two basic experimental approaches. The first involves recording the electrical activity of active units, and the second is similar to the glycogen depletion process described in Chapter 4.

### Electrical activity

Information about the order in which motor units are recruited in human muscle can be obtained by recording the electromyographic (EMG) activity from single motor units using fine wire or needle electrodes inserted into the muscle. As the subject makes a steady contraction it is usually possible, by manipulating the electrode, to find a position where a dominant persistent spike is seen (Fig. 5.1). This spike comes from the muscle fibre closest to the tip of the electrode and gives an indication of the firing frequency of the motor unit to which that particular fibre belongs. As the electrode is moved about in the muscle, fibres from other

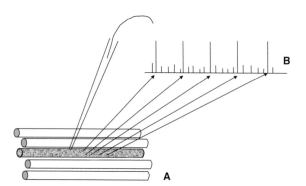

**Figure 5.1** Recording single motor unit activity. **A**, needle electrode with tip close to one fibre (shaded). **B**, electrical activity is recorded from a number of fibres but with one shaded fibre giving a large regular spike.

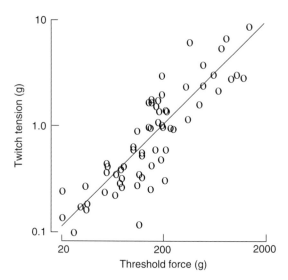

**Figure 5.2** Twitch forces produced by single motor units in one subject as a function of the force at which the motor units were recruited. Force in this case is reported in grams weight, approximately 0.01 N. (Redrawn from the data of Milner-Brown et al 1973.)

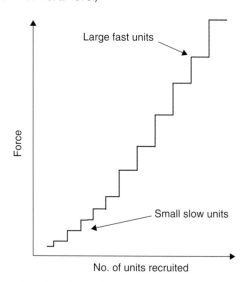

**Figure 5.3** Regulation of force by recruitment. Small motor units are recruited for low-force contractions, larger units being required for higher forces.

motor units can be identified and their pattern of firing examined.

Motor units become active at characteristic levels of voluntary force. The force contribution of a single unit will be superimposed on force generated by other active units, which are firing randomly with respect to the unit being recorded. The EMG signal from the single unit can be used to trigger a signal averager, so that the contribution of force from the one unit can be identified and the size and contractile characteristics (speed) of the twitch of that motor unit measured.

Motor units recruited during low-force contractions prove to be small with relatively slow twitches, whereas the units recruited when making high-force contractions are larger, fast units (Fig. 5.2). This pattern of recruitment is consistent with there being a hierarchy of motor units, with small, slow, motor units being active during low-force contractions while the fast units are active only during higher-force contractions. Henneman and co-workers found in the cat that small, slow, motor units were supplied by small, easily excitable motoneurons, while larger units were innervated by motoneurons that had higher thresholds for excitation. Henneman suggested that this difference might be the basis for the modulation of force, with units being recruited in order of their size, and this idea has become known as *Henneman's size principle* (Henneman et al 1974). By progressive recruitment of motor units, the force generated by a muscle can be increased in a stepwise fashion (Fig. 5.3).

The smaller, easily recruited, motor units are also the most fatigue resistant (see Ch. 4), and the recruitment pattern described above has an obvious advantage in that the most frequently used units are small, slow and fatigue resistant. Thus, they can provide enough force for the majority of

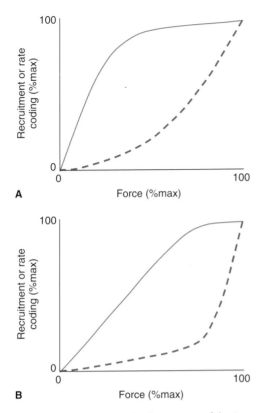

**Figure 5.4** Model of the relative importance of the two strategies, recruitment and rate coding, for contractions of different strengths. Solid lines indicate the recruitment of motor units, dashed lines the involvement of rate coding. **A**, a small distal muscle (e.g. in the hand) where, over most of the range of force, rate coding plays a major part. **B**, a large proximal muscle (e.g. quadriceps or biceps) where recruitment is the major way of modulating force.

everyday activities such as postural adjustments and walking, which require relatively small forces over prolonged periods. The large, fast and rapidly fatiguable units are used only for high-force contractions such as sprinting or jumping, which usually happen only a few times a day, giving plenty of time for recovery.

If this were the only mechanism for modulating force, the steps in force would be small at low forces (i.e. recruiting small motor units), while at higher forces the increments would be larger, because the recruited motor units are larger (Fig. 5.3). Consequently at higher forces, control would be expected to be less precise and, to some extent, this is what is seen. However, in addition, force

can also be modulated by varying the frequency of stimulation, and this is known as 'rate' or 'frequency' coding. The force generated by a muscle depends on the frequency of activation, and muscle fibres of different speeds have characteristic force–frequency relationships (see Figs 4.6 & 4.7). There is scope for considerable control of force if the motoneuron firing frequency changes in the range corresponding to the steep part of the force–frequency curve. Motoneurons have an intrinsic firing frequency which can be regulated up or down by inhibitory and stimulatory inputs from the brain and by afferent signals from chemoreceptors and mechanoreceptors in the skin, muscle, tendons and joints.

In reality, both recruitment and rate coding are used to control muscle force during voluntary movements. In general, recruitment of motor units is used to modulate lower forces, whereas for larger contractions rate coding becomes more important. The extent to which these two strategies are used varies with different muscles; recruitment tend to be the main mechanism with the larger proximal muscles, while smaller distal muscles involved in fine movements make more use of frequency coding (Fig. 5.4). As examples, in the biceps brachii, recruitment of new motor units occurs up to approximately 90% of the maximal force, while for the adductor pollicis all motor units are recruited by 50% maximal force and the additional force is obtained by rate coding (Kukulka & Clamann 1981). Most motor units are recruited at frequencies of 8–12 Hz, while the maximal frequencies during steady high-force contractions may be 20–50 Hz (Bellemare et al 1983, De Luca et al 1982). An advantage of having a large element of rate coding is that it allows very precise control of force, which is a desirable feature for small muscles such as those in the hand. The modulation of force primarily by recruitment in large proximal muscles gives a cruder control, but these muscles are generally not required to perform delicate movements.

The recruitment of motor units and the frequency with which they are activated is determined by the excitability of the motoneuron innervating these fibres. In general, the pattern of recruitment and firing is relatively constant and characteristic of a given motoneuron and motor unit. However, there are situations where motoneuron firing rates can change. When it is necessary to generate forces rapidly, such as during ballistic movements, the motoneurons can discharge initially at a much higher rate, for example in the region of 150 Hz for two or three impulses. The effect of this is to bring the motor unit force rapidly to a tetanic level, where it can then be maintained at a lower frequency of activation.

## Metabolic assessment of recruitment

The broad principles of motor unit recruitment have been established by the type of EMG studies described above. For technical reasons, most electrical studies are limited to superficial muscle groups (frequently distal muscles in the hand or lower leg) and restricted to isometric contractions. It is clearly also of interest to know how recruitment patterns vary in the major muscle groups during the course of whole-body exercise of different intensities, such as cycling or running at different speeds where the muscles are both shortening and being lengthened.

The approach that has been used in these situations is to take muscle biopsy samples after exercise, stain them for glycogen and assess the extent to which different fibre types have become depleted. Figure 5.5 shows the results of a number of experiments that studied the effects of different intensities and durations of cycle exercise. With a low exercise intensity (approximately 30% $V_{O_{2max}}$), type 2 fibres were hardly recruited in the first 2 h of exercise. With increasing intensity, type 2a fibres played an increasing role, and at 75% $V_{O_{2max}}$ all fibre types were involved at some time during 2 h of exercise.

There are a number of limitations to this methodology. The first is that histochemical staining for glycogen gives only a qualitative measure of depletion, because it is not possible to measure the content of the same fibres before and after exercise, and a relatively long duration of exercise is required for there to be significant depletion. Resting glycogen levels also vary with fibre type, which makes assessment of depletion more difficult. Furthermore, although glycogenolysis is

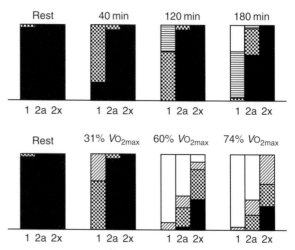

**Figure 5.5** Fibre recruitment during exercise with different intensities and duration. Schematic illustration of glycogen content in different fibre types (1, 2a and 2x) of human quadriceps muscle after exercise. Glycogen content of fibres is based on subjective rating (dark, filled with glycogen; white, depleted; others in between) and the proportion of fibres with different glycogen levels presented. Upper panel, the effect of exercise duration (at 31% $V_{O_{2max}}$) for the different fibre types. Lower panel, the effect of intensity of exercise (duration 120 min). (Redrawn from the data of Saltin & Gollnick 1983.)

activated by intracellular calcium concentration, it is also stimulated by adrenaline (see Ch. 6), so that depletion may not necessarily indicate muscle activity. Different fibre types have different rates of glycogenolysis, which will depend in part on the extent of oxidative metabolism and the use of alternative substrates such as lipid. Consequently a greater depletion may simply indicate different substrate utilization rather than different recruitment.

An alternative approach to the study of recruitment patterns during exercise of relatively short duration is to measure some other metabolites that change with activity, the most obvious being phosphocreatine, creatine and phosphate (Beltman et al 2001). Measurement of the ratio of creatine phosphate to creatine in fragments of single fibres may be a way forward, although again it should be remembered that different fibre types have different ratios, even at rest, and the analysis of single-fibre fragments is very labour intensive.

## SENSORY RECEPTORS IN SKELETAL MUSCLE

There are a variety of sensory receptors in muscle. Some of these signal chemical changes in the muscle; others respond to changes in pressure, force and muscle length.

### Muscle spindles

Spindles are found scattered throughout a muscle, some in the belly and others near the musculo-tendinous junctions. Each spindle consists of a number of highly specialized muscle fibres contained within a collagenous capsule; these fibres are described as *intrafusal* fibres, as opposed to the majority of muscle fibres with which we are familiar, which are *extrafusal*. The important feature of the spindle fibres is that they run in parallel with the extrafusal fibres and are thus subjected to length changes as the muscle shortens or is stretched. A diagrammatic representation of a muscle spindle in longitudinal section is shown in Figure 5.6. There are two types of intrafusal fibre, the larger fibres are about 25 μm in diameter and the smaller fibres are less than 20 μm, both being less than half the diameter of the extrafusal fibres. The length of a spindle in a cat muscle is about 7–10 mm. In the central region of the larger intrafusal fibres there is a cluster (bag) of 50–100 nuclei; fibres of this type have been designated *bag fibres*. The nuclei of the smaller extrafusal fibres form a central chain, and these fibres have come to be known as *chain fibres*. These central regions of the intrafusal fibres do not contain contractile material.

### Afferent innervation of spindles

The large, myelinated Ia primary afferent (i.e. conducting from the periphery to the central nervous system) nerve fibres have spiral endings around the central portion of each bag and chain fibre. There are also secondary endings of the smaller, myelinated, type II afferents present and these are located mainly at the ends of the chain fibres.

### Response to stretch

In response to a ramp and hold stretch there is a rapid increase in the frequency of firing in the primary Ia afferents originating on the bag fibres (Fig. 5.7A). During the hold phase, when the spindle is at a constant length, the rate of firing decreases. Bag fibres are said to accommodate to the stretch and therefore act mainly as indicators of rate of change.

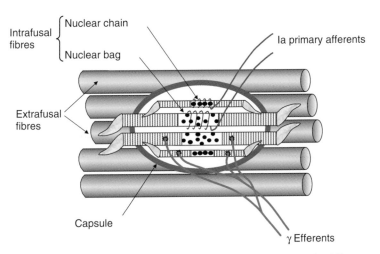

**Figure 5.6**  Diagram showing the main features of a muscle spindle. Afferent innervation is shown only for the upper chain and bag fibres, and efferent innervation is shown for the lower fibres. Secondary afferents innervate the ends of the chain fibres, but these are not shown here for the sake of clarity.

The chain fibres behave in a more *elastic fashion* and give a sustained high-frequency response during the hold phase (Fig. 5.7B) and therefore indicate muscle length. This signal is mainly from the secondary endings (type II afferents), but there are also primary afferents on the chain fibres so that the signal in the type Ia afferents may be a mixture derived from the responses from both types of intrafusal fibre.

### Efferent innervation of spindles

The efferent innervation (i.e. conducting outwards, from the central nervous system to the periphery) comes from γ motoneurons (sometimes known as fusimotor neurons), axonal branches of which form synapses with both types of intrafusal fibre on either side of the central nucleated portion. The chain and bag fibres are similar to extrafusal fibres in the arrangement

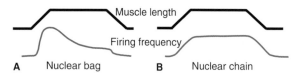

**A**    Nuclear bag       **B**    Nuclear chain

**Figure 5.7** Response of intrafusal fibres to stretch. Bag fibres accommodate to stretch (**A**), whereas the chain fibres remain stretched and continue to fire (**B**).

and function of their contractile proteins, except that they do not extend through the central portions of the fibres and the contractions are localized to regions on either side of this. When the γ motoneurons cause the spindle fibres to contract, the effect is therefore to stretch the central portion of the intrafusal fibres around which the sensory nerve endings are situated (Fig. 5.8).

When the γ motoneuron fires, the two ends of each intrafusal fibre contract, stretching the central portion and activating the primary and secondary afferent endings. The afferent activity stimulates the α motoneurons, causing the extrafusal fibres to contract; when these fibres shorten, the stretch on the intrafusal fibres is removed and the afferent signal is reduced (Fig. 5.8C).

### Input to the central nervous system

The spindle afferents have their cell bodies in the dorsal root ganglia and have excitatory synaptic inputs to the α motoneurons supplying the muscle in which the spindles are situated.

Spindles function as stretch receptors in a servo loop controlling length in a muscle. Stretch of the spindle causes afferent discharge, which activates the α motoneurons producing contraction of the muscle; an example of this is the simple stretch

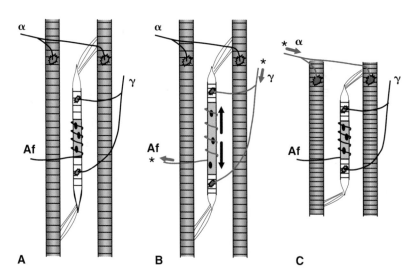

**Figure 5.8** Action of muscle spindles. **A**, starting position; **B**, γ activity (*) stretches the central portion of the spindle (arrows) causing afferent discharge (*); **C**, α-motoneuron activity (*) causes contraction of extrafusal fibres, shortening of the spindle and reduced afferent discharge. Af, afferent fibres (type Ia or II).

reflex (Fig. 5.9A). Not only do the spindle afferents feed back to the α motoneurons of the muscle in which they are situated, they also have a synaptic input to the α motoneurons of antagonist muscles, where they exert an inhibitory influence (Fig. 5.9B). This inhibition ensures that antagonist muscles do not work against one another. Alternatively, activity in the γ motoneurons can set the muscle length: the higher the activity, the shorter the length.

In addition to the afferent input from the muscle spindles, there is an inhibitory input from Golgi tendon organs, which act via 1b inhibitory interneurons, as well as a feedback inhibition from Renshaw cells. Not shown in Figure 5.9 are the inhibitory inputs to the α motoneurons arising from mechanoreceptors and chemoreceptors in muscle as well as in skin, joints and ligaments.

The extent to which normal movements are controlled by the activity of α or γ motoneurons is not known. There have been suggestions that small, low-force, postural movements may involve activation of the γ motoneurons, whereas rapid movements are the result of direct descending activation of α motoneurons. Alternatively, or in other situations, there may be co-activation of α and γ motoneurons. Activity of the γ motoneurons will ensure that the spindles do not get left behind when the rest of the muscle shortens, and so can still remain responsive to stretch, increasing the sensitivity of the servo loop and thereby giving greater control during a movement.

## REFLEX ACTIVITY

Reflex activity, which is the involuntary contraction of skeletal muscle in response to a stimulus, may either traverse a relatively short spinal pathway or take a longer path via the higher centres in the brain. These two kinds of reflex can be identified by their latency, or the time taken for a signal to traverse the reflex arc. For spinal reflexes this time is of the order of 10–20 ms, while for supraspinal reflexes it is 50 ms or more.

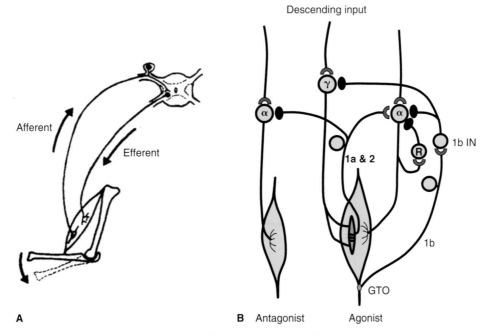

**Figure 5.9** Spinal pathways regulating muscle contraction. **A**, simple reflex; afferent activity activates the α motoneuron causing contraction of the extrafusal muscle fibres. **B**, more complex wiring diagram of α and γ innervation and the afferent inputs modifying their activity. Y-shaped endings indicate excitatory synapses; solid endings show inhibitory synapses. GTO, Golgi tendon organ; R, Renshaw cell; 1b IN, inhibitory interneuron.

## Spinal reflexes

Spinal reflexes play an important role in coord-inating the activity of local groups of muscles such as those in the hand, where a variety of excitatory and inhibitory reflexes arise from the skin, joints and muscles regulating such delicate actions as gripping without crushing.

## Supraspinal reflexes

Supraspinal reflexes are chiefly involved in the coordination of muscular activity in various parts of the body where, for instance, this may be neces-sary for the maintenance of balance. This can be illustrated if a small stretch is applied to the flexor of the thumb. A spinal reflex can be observed in the flexor muscle, but there are also supraspinal reflexes of long latency in a wide range of muscles on the contralateral side of the body. The impor-tance of these muscular contractions is to main-tain balance so that the subject does not fall as the result of the pull on his or her thumb. With-out these compensating contractions, even the simplest movement such as lifting a cup would unbalance the body. As humans have a highly unstable form of locomotion, the maintenance of posture and balance constitutes a most important role for these reflexes.

It can be seen that the spindles play an import-ant role in the fine control of muscle action. It is not surprising, therefore, to find that muscles of the neck and hand have a very high density of spindles—approximately ten times the density found in the larger locomotory muscles.

## Golgi tendon organs

Tendon organs are formed by encapsulated end-ings of large myelinated axons, known as Ib affer-ent fibres, situated at musculotendinous junctions. They are concerned with providing information about muscle force to the central nervous system. The information from Golgi tendon organs acts via an inhibitory interneuron to reduce the excit-ability of $\alpha$ and $\gamma$ motoneurons (Fig. 5.9B). In the cat soleus there are about 50 Golgi tendon organs, and each tendon organ is responsive to a group of muscle fibres closely connected with that portion of tendon. Contraction of only one or two of these fibres can make the tendon organ discharge. Because the fibres of individual motor units are scattered in the muscle, it may be any one of a number of motor units that drives the tendon organ at any one time.

## Other afferent inputs

The Ia and II afferent inputs from the spindles are important in controlling $\alpha$ motoneuron activ-ity, exerting a strong excitatory influence. There are a number of other modulating influences, including Renshaw cells (small interneurons in the spinal cord that receive collateral input) and descending inhibitory and excitatory pathways affecting the excitability of the $\alpha$ motoneurons (Fig. 5.9B).

Information also comes to the central nervous system from pressure receptors in the skin and nociceptors in the muscle and fascial sheath, which respond to metabolite changes in the active mus-cle and may exert an inhibitory influence on $\alpha$ motoneurons.

## COORDINATION OF MUSCULAR CONTRACTIONS

All bodily movements, while usually seeming quite simple for a healthy adult, nevertheless involve highly complex patterns of muscular con-traction. These movements are generally made without any consciousness of the complexity or even the contraction sequence of the various muscles used. For a task such as reaching out to pick up a cup, the simplest sequence of muscle commands must include:

1. activate elbow extensors—arm extends
2. activate finger extensors—hand opens
3. activate finger flexors—cup is grasped
4. activate elbow flexors—cup is brought up towards body.

For a very young baby, this would represent a major achievement requiring great concentra-tion in order to activate the right muscles in the correct sequence. With time and practice the conscious command 'pick up that cup' serves to initiate the correct sequence of movements without conscious effort. This problem may be thought of in terms of robotics or computer control in which the sequence 1–4 above would be stored as a subroutine to be called up whenever the appropriate movement was required. Nearly every action in the daily life of an adult is encoded in this way and it is only when we encounter an unfamiliar activity such as writing with our non-dominant hand, or learning to ski or to ride a bicycle as an adult, that we become conscious of the complex learning process involved. We also experience the grim determination and concentration that is required to learn the new task.

It is of particular interest to know where in the central nervous system these subroutines are located. In insects, various locomotor and other functions have been localized to specific neuronal groups. In higher animals, locomotor activity is integrated with other bodily activities and it is not easy to identify discrete locomotor centres. There are, however, some situations where activities or movements can be demonstrated to be under the control of the spinal cord and independent of the higher motor centres. A decapitated chicken will continue to flap its wings for some time, and complex locomotor movements can be demonstrated in decerebrate cats and dogs if the body is supported over a moving treadmill. Locomotion in cats and dogs is a complex task involving the coordination of many muscle groups in four separate limbs and, moreover, the pattern of locomotion will change in response to changes in treadmill speed. The computing power to achieve this must be considerable and must exist within the spinal cord.

Spinal cord-injured patients, however, do not show automatic locomotor reflexes; this indicates that human locomotion requires some input from the higher centres, although this input need not necessarily be conscious. It may also reflect the fact that an upright posture is much more demanding in terms of coordination than four-legged locomotion, and balancing an unstable body on two legs requires the greater computing power available in the higher motor centres—just as modern high-performance aircraft are inherently unstable and require sophisticated computer control to keep them flying.

It may still, however, be oversimplistic to think of the movement subroutines in terms of specific instructions concerning individual muscles. This is suggested by an observation described by Merton (1972). He noticed that when he wrote his signature in very small letters using a magnifying glass and a mapping pen it was recognizably the same as when he wrote it large on a wall with a paint brush. The specific muscles used to perform these two tasks were very different: in the first case, small delicate movements by the hand muscles and, in the second, large bold movements of the whole arm and shoulder. Despite this difference, the signature was characteristically his own in both cases, suggesting that the final control is situated in some higher central area and exists in a much more complex form than a mere sequence of muscle movements.

## REFERENCES AND FURTHER READING

Armstrong D M 1988 The supraspinal control of mammalian locomotion. Journal of Physiology 405: 1–38

Bellemare F, Woods J J, Johansson R, Bigland-Ritchie B 1983 Motor-unit discharge rates in maximal voluntary contractions of three human muscles. Journal of Neurophysiology 50: 1380–1392

Beltman M, van Mechelen W, Sargeant A J, de Haan A 2001 Changes in the PCr/Cr ratio: a measure for human fibre activation during short-term maximal voluntary exercise. Journal of Physiology 533: 128P

Boyd I A 1976 The mechanical properties of dynamic nuclear bag fibres, static nuclear bag fibres and nuclear chain fibres in isolated cat muscle spindles. Progress in Brain Research 44: 33–50

Burke R E, Levine D M, Tsairis P, Aajai F E 1973 Physiological types and histological profiles in motor units of cat gastrocnemius. Journal of Physiology 234: 723–748

Burke R E, Levine D M, Zajac F E, Tsairis P, Engel W K 1971 Mammalian motor units: physiological–histochemical correlation in three types in cat gastrocnemius. Science 174: 709–712

De Luca C J, LeFever R S, McCue M P, Xenakis A P 1982 Behaviour of human motor units in different muscles during linearly varying contractions. Journal of Physiology 329: 113–128

Fournier E, Pierrot-Deseilligny E 1989 Changes in transmission in some reflex pathways during movement in humans. News in Physiological Sciences 4: 29–32

Garnett R A F, O'Donovan M J, Stephens J A, Taylor A 1979 Motor unit organization of human medial gastrocnemius. Journal of Physiology 287: 33–43

Henneman E, Clamann H P, Gillies J D, Skinner R D 1974 Rank order of motoneurons within a pool, law of combination. Journal of Neurophysiology 37: 1338–1349

Kugelberg E 1973 Histochemical composition, contraction speed and fatiguability of rat soleus motor units. Journal of the Neurological Sciences 20: 177–198

Kukulka C G, Clamann H P 1981 Comparison of the recruitment and discharge properties of the motor units of human brachial biceps and adductor pollicis during isometric contractions. Brain Research 219: 45–55

Marsden C D, Merton P A, Morton H B 1976 Stretch reflex and servo action in a variety of human muscles. Journal of Physiology 259: 531–560

Marsden C D, Merton P A, Morton H B 1977 The sensory mechanism of servo action in human muscle. Journal of Physiology 265: 521–535

Merton P A 1972 How we control the contraction of our muscles. Scientific American 226: 30–37

Milner-Brown H S, Stein R B, Yemm R 1973 The orderly recruitment of human motor units during voluntary isometric contractions. Journal of Physiology 230: 359–370

Peter J B, Barnard V R, Edgerton V R, Gillespie C A, Stempel K E 1972 Metabolic profiles of three fibre types of skeletal muscles in guinea pigs and rabbits. Biochemistry 11: 2627–2633

Ranvier M L, 1873 Propriétés et structures différentes des muscles rouges et des muscles blancs, chez les Lapins et chez les Raies. Comptes Rendus des Académie de Sciences 77: 1030–1034

Saltin B, Gollnick P D 1983 Skeletal muscle adaptability: significance for metabolism and performance. In: Peachey L D, Adrian R H, Geiger S R (eds) Handbook of physiology, section 10: skeletal muscle, pp 540–555. Bethesda, American Physiological Society

Yemm R 1977 The orderly recruitment of motor units of the masseter and temporal muscles during voluntary isometric contractions. Journal of Physiology 265: 163–174

## QUESTIONS TO THINK ABOUT

1. What is the *size principle* of Henneman et al (1974) and what are the advantages and possible disadvantages of recruitment according to this principle?

2. Explain how recruitment and rate coding are used in controlling force. Do you expect a difference in the relative importance of these two methods between fast and slow, and between large and small muscles?

3. Glycogen depletion can be used to identify motor units that have been recruited. How do you explain the fact that following electrical stimulation when all fibres in a muscle have been activated there are differences in glycogen depletion in different fibre types?

4. Figure 5.5 shows the patterns of glycogen depletion with different times and intensities of exercise. Why is it that the type 1 fibres, which have the lowest metabolic rate and highest ability to oxidize other substrates such as fat, are still the most depleted fibre type?

5. What are the limitations of using glycogen depletion as a marker for recruitment, and are there any other metabolites that could be used in this way?

6. What are the differences between intrafusal and extrafusal muscle fibres, and between efferent and afferent nerves?

7. How could stimulation of $\gamma$ efferents be used to control the length of a muscle? In these situations, would the nuclear bag or chain fibres be providing the relevant afferent information?

8. What are the major excitatory and inhibitory inputs to the $\alpha$ motoneuron?

9. What is the function of the Golgi tendon organs and could they play a role in protecting muscle and tendon?

10. How would you distinguish a spinal from a supraspinal reflex?

# 6

# Energy metabolism during exercise

In preceding chapters we have dealt with the machinery and control of muscle activity but little or no reference has been made to the source of the energy that drives these processes. All living things need energy to move, to grow or simply to maintain the status quo. Ultimately this energy comes from the sun or, in a few rare cases, from hot springs deep in the ocean, but heat or sunlight energy cannot directly power the processes of life; an intermediate is required, and this is adenosine triphosphate (ATP), often described as a 'high energy' compound. Hydrolysis of the terminal phosphate is an exothermic process yielding energy (Fig. 6.1A), and the value of ATP to living cells is that if the terminal phosphate is transferred to a protein or carbohydrate the energy, or at least part of it, is imparted to that molecule. A phosphorylated protein may change its configuration and move (e.g. transport proteins in the cell membrane or myosin in muscle), or a carbohydrate may be primed to begin a sequence of reactions yielding even more energy (e.g. sugar phosphates in the glycolytic pathway). This subject has been the essence of biochemistry for the past 100 years and, as this chapter gives only an outline of the pathways and regulatory mechanisms, the reader is referred to standard biochemistry textbooks for the details.

ATP is the currency of energy and its ubiquitous presence in all organisms, including bacteria, plants and animals, clearly testifies to its importance in living cells. Curiously, for such an important substance, it is found in relatively low concentrations within cells ($2–8 \, mmol \, L^{-1}$), probably because it is such a reactive substance that at

**A  ATP hydrolysis**

$$ATP + H_2O \rightleftharpoons ADP + H^+ + P_i + \text{'energy'}$$

**B  Creatine kinase reaction**

$$PCr + ADP + H^+ \rightleftharpoons Cr + ATP$$

**C  Adenylate kinase reaction (a) and myoadenylate deaminase reaction (b)**

$$a.\ ADP + ADP \rightleftharpoons ATP + AMP$$

$$b.\ AMP \longrightarrow IMP + NH_3$$

**D  Anaerobic glycolysis**

$$Glucose + 2\ ADP + 2\ P_i \longrightarrow 2\ lactate + 2\ ATP + 2\ H_2O$$

$$Glycosyl\ unit + 3\ ADP + 3\ P_i \longrightarrow 2\ lactate + 3\ ATP + 3\ H_2O$$

**Figure 6.1**  Reactions of ATP hydrolysis and anaerobic regeneration. ADP, adenosine diphosphate; AMP, adenosine monophosphate; ATP, adenosine triphosphate; Cr, creatine; IMP, inosine monophosphate; $P_i$, inorganic phosphate; PCr, phosphocreatine.

high concentrations it would cause chaos, reacting indiscriminately with cell components. For tissues that have a steady and slow metabolic rate, a low ATP content may be no problem with supply easily keeping up with demand. Skeletal muscle, however, can increase its metabolic rate 60–100-fold during the transition from rest to activity and would use up the available ATP within a matter of seconds if it were not regenerated by pathways that can respond rapidly.

# REGENERATION OF ATP

There are several pathways that regenerate ATP, some of which are very rapid (Fig. 6.1) but can supply only limited amounts of ATP, others are slower but of much higher capacity.

## Creatine kinase reaction

Creatine (Fig. 6.2) is synthesized from three amino acids in the liver (although it is also obtained by

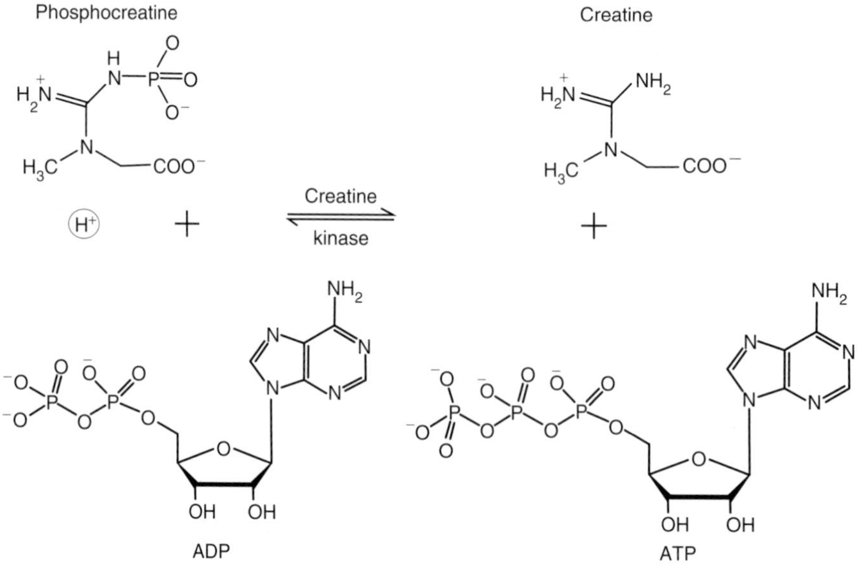

**Figure 6.2**  Creatine kinase reaction. Note that a proton is taken up when ATP is synthesized.

eating meat) and is taken up by muscle from the circulation. Creatine is phosphorylated in a reaction catalysed by the enzyme *creatine kinase* at the expense of ATP. Under physiological conditions, the free energies of phosphocreatine (PCr) and ATP hydrolysis are similar, so that the forward and back reactions (Fig. 6.2) are very sensitive to changes in concentrations of the reactants. There is also a very high concentration of creatine kinase in muscle fibres and the interconversion of PCr and ATP is extremely rapid. At rest, about 80% of the muscle fibre creatine content is present as PCr and this is at a concentration about five times greater than ATP itself. In invertebrate muscle the role of creatine is taken by arginine, with arginine phosphate being the immediate buffer of the ATP stores.

## Adenylate kinase reaction

Another way of rapidly forming small amounts of ATP is to transfer phosphate from one molecule of ADP to another, generating, in the process, one molecule of ATP and one of adenosine monophosphate (AMP). This reaction is coupled to the deamination of AMP (catalysed by AMP deaminase), which removes the AMP from the equilibrium and drives the reaction to the right (Fig. 6.1C). Quantitatively this process is not very important and tends to occur when the reserves of PCr are exhausted. The main purpose of the reaction is to remove ADP, the accumulation of which tends to inhibit reactions involving ATP. A consequence of this process is that once AMP is deaminated (within seconds) the resynthesis of adenine nucleotides is relatively slow (several minutes) through the purine nucleotide cycle.

## Glycogenolysis and glycolysis

Sunlight energy is used by plants to make hexose sugars from carbon dioxide and water in the photosynthetic process. Animals obtain their energy by reversing this process, degrading the sugars back to carbon dioxide and water, and storing the liberated energy as ATP. The process is a complex one involving many steps, starting with the anaerobic glycolytic pathway in which a six-carbon sugar is broken down to two three-carbon

fragments, followed by an aerobic process in which the three-carbon fragments are taken apart to produce $CO_2$ and the associated hydrogens are oxidized to give water.

### Breakdown of glucose and glycogen

During high-intensity exercise most of the energy from carbohydrate comes from glucose stored as intramuscular glycogen. The first stage in obtaining energy from glycogen is the sequential hydrolysis of glycosyl units from the ends of the chains, in the course of which phosphate is attached to form glucose 1-phosphate (Fig. 6.3). The enzyme catalysing this process is *phosphorylase,* which is one of the key enzymes in glycogenolysis. Phosphorylase hydrolyses 1–4 linkages of glycogen but cannot deal with the branch points formed by the 1–6 bond and leaves a short side-chain of four glycosyl units. The *debranching enzyme* then moves three of these units and attaches them to the end of an adjacent long chain where they can again be attacked by phosphorylase, yielding glucose 1-P. The remaining unit is removed by *amyloglucosidase,* which hydrolyses the 1–6 bond to give free glucose.

Glucose 1-P is converted first to glucose 6-P and then to fructose 6-P. More energy is required to convert the fructose 1-P to fructose 1,6-biphosphate; the enzyme that does this, *phosphofructokinase,* is an important regulatory step in glycolysis. By this stage the six-carbon sugar has been phosphorylated twice and has a symmetrical structure. It also has sufficient energy to take it over the energy barrier so that, by the action of *aldolase,* the six-carbon ring is split into two three-carbon fragments. From that point a series of dehydrogenase and substrate level phosphorylation reactions ultimately yields two molecules of pyruvate and four of ATP for every glycosyl moiety entering the glycolytic pathway. The reactions involved are summarized in Figure 6.3.

### Energy yield of glycolysis

The precise energy balance of glycolysis depends on the source of the carbohydrate. If the glucose comes from the blood, it has to be phosphorylated

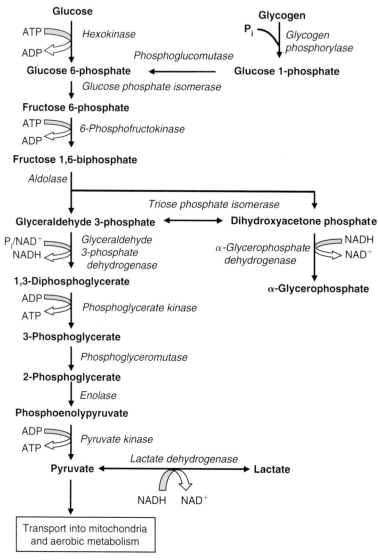

**Figure 6.3**   Overview of glycolysis, including the phosphorylase step and the α-glycerophosphate reaction. Enzyme names appear in italics.

twice before getting to the stage where energy can be obtained, and there is a net gain of one ATP per mole of pyruvate formed. If the glycosyl moiety has come from the hydrolysis of glycogen (the most common situation) then only one ATP is required to form fructose 1,6-biphosphate, as the other phosphate comes from inorganic phosphate during the phosphorylase reaction. Consequently, one ATP is used and four are generated, giving a net yield of 1.5 ATP per pyruvate. It must be

realized that, overall, obtaining energy from glycogen is no more efficient than metabolizing glucose, because the formation of glycogen is itself an energy-requiring process. Glucose taken up into the muscle is phosphorylated and then activated by combining with uridine triphosphate to form UDP–glucose (releasing pyrophosphate in the process) before condensing to form the giant branching polymer structure that is glycogen. The formation of glycogen occurs over a relatively long time course so that during rest periods energy is stored in the glycosyl bonds in addition to that present in the basic carbohydrate skeleton.

### Importance of NAD

The final energy-yielding part of glycolysis depends critically on the dehydrogenase reaction in which electrons are removed from glyceraldehyde 3-phosphate and accepted by the coenzyme nicotinamide–adenine dinucleotide (NAD) to form NADH. The supply of NAD is limited and, if glycolysis is to continue for more than a second or two, the NAD has to be regenerated. During 'aerobic' activity the NAD is regenerated by oxidation in the mitochondria; there are three ways in which this can occur. The first is in a reaction that converts oxaloacetate to malate—the malate shuttle. The malate formed is transported into the mitochondrion where it is converted back to oxaloacetate, generating NADH, which can be oxidized by the electron transport chain. The second is by the conversion of dihydroxyacetone phosphate into α-glycerophosphate, which passes into the mitochondrion where it is dehydrogenated to generate the reduced form of flavin–adenine dinucleotide (FADH$_2$). This FADH$_2$ is oxidized by the electron transport chain while the dihydroxyacetone phosphate diffuses back into the cytoplasm. For moderate intensities of exercise these two shuttles keep pace with the requirement of glycolysis for NAD. If, however, the workload is high, or the muscle is forced to work under anaerobic conditions, α-glycerophosphate accumulates (see Fig. 6.7B) because it can no longer be oxidized in the mitochondria, and a third regeneration reaction comes into play. NAD is regenerated by converting pyruvic acid to lactic acid, catalysed by *lactate dehydrogenase*. Lactic acid is, therefore, not so much the end-product of glycolysis but a way of regenerating NAD and allowing glycolysis to continue for as long as possible (Fig. 6.3).

Although appearing complex when written out as in Figure 6.3, the glycolytic reactions are rapid, so the rate of ATP regeneration from this source is about one-half to one-quarter that of the creatine kinase reaction. Although large quantities of glycogen are stored in muscle fibres, there is a

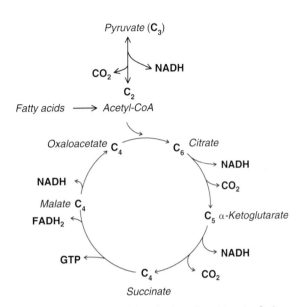

**Figure 6.4** Overview of the tricarboxylic acid cycle. CoA, coenzyme A; FADH$_2$, reduced form of flavin–adenine dinucleotide; GTP, guanosine triphosphate; NADH, reduced form of nicotinamide–adenine dinucleotide.

limit to the extent of glycolysis, which is set by the accumulation of lactate and the acidification of the muscle fibre. Anaerobic glycolysis can yield about the same quantity of ATP as that obtained from the creatine kinase reaction (i.e. about 40 mmol per kg wet weight of muscle).

## Pyruvate and acetylcoenzyme A

Glycolysis fulfils two functions. It generates a limited, although important, amount of energy during anaerobic or high-intensity exercise, and also provides pyruvate as a substrate for oxidative metabolism. Glycolysis occurs in the cytoplasm of the cell, while the further metabolism of pyruvate occurs inside the mitochondria.

Inside the mitochondria, pyruvate, a three-carbon molecule, is first decarboxylated to a two-carbon fragment before entering the tricarboxylic acid (TCA) cycle (Fig. 6.4). This conversion is catalysed by a series of enzymes, known as the *pyruvate dehydrogenase complex* (PDH). The net result of this activity is that pyruvate is decarboxylated and bound to coenzyme A (CoA), forming acetyl-CoA and, in the process, NADH is generated.

Acetyl-CoA is the common entry point of carbohydrate, fat and amino acids into the TCA cycle. The mobilization and oxidation of fatty acids is a major source of energy during prolonged exercise, with β oxidation producing acetyl-CoA units that are fed into the TCA cycle.

## Tricarboxylic acid cycle

The first step in the cycle is the condensation of acetyl-CoA with oxaloacetic acid to form citric acid and, in the process, liberating coenzyme A. Thereafter, there are a series of dehydration and decarboxylation reactions during which carbon dioxide and reducing equivalents in the form of NADH and $FADH_2$ are produced together with a small quantity of guanosine triphosphate (GTP), eventually ending at the starting point of the cycle, oxaloacetate (Fig. 6.4). The cycle takes place within the matrix of the mitochondria.

Much of exercise physiology involves the measurement of oxygen consumption and carbon dioxide production, and it is often implied that oxygen is used to produce the carbon dioxide. In fact, atmospheric oxygen is used only to combine with protons in the electron transport chain and does not produce carbon dioxide, which comes from the decarboxylation reactions of PDH and the TCA cycle (Fig. 6.4).

## Electron transport chain

Although small amounts of energy are produced by substrate level phosphorylation in the TCA cycle, the most important function of this series of reactions is to generate reduced coenzymes that are oxidized by the electron transport chain to yield water and a large quantity of ATP.

The electron transport chain consists of a series of enzyme complexes and cytochromes on the inner mitochondrial membrane (Fig. 6.5). Protons and electrons are separated from the reduced coenzymes NADH and $FADH_2$, and the energy obtained from the electrons passing down the electrochemical gradient of the electron transport chain is used to transport protons against a concentration gradient from the inner to the outer surface of the inner mitochondrial membrane. Protons then pass back into the interior of the mitochondrion through a pore in complex V, and the energy released in this process is used to synthesize ATP from ADP and $P_i$. There are three sites at which protons are transported from the interior of the mitochondrion, at complexes I, III and IV. NADH donates an electron and $H^+$ at complex I (NADH dehydrogenase); these electrons drive the transport of protons at all three sites as they pass down the electrochemical gradient and are thus associated with the synthesis of 3 ATP. $FADH_2$, however, donates electrons at complex II (succinate dehydrogenase), and these electrons are responsible for the synthesis of only 2 ATP. The flow of electrons down the electron transport chain can continue only while oxygen is present to act as the final electron acceptor, and the two negatively charged oxygen atoms formed from molecular oxygen react rapidly with protons to form water. The electron transport chain thus deals with reduction equivalents arising from the TCA cycle together with NADH generated in the cytoplasm during glycolysis.

The oxidative metabolism of carbohydrate is clearly a complex process and the production of ATP is relatively slow when compared with that from glycolysis or phosphocreatine. It is, however, an efficient use of an energy store, as the complete aerobic metabolism of 1 molecule of glucose yields 38 ATP molecules, which is nearly 20 times more than the production of 2 molecules of ATP from glycolysis alone.

## CONTROL OF METABOLIC PATHWAYS

In a tissue such as skeletal muscle with widely fluctuating levels of activity and metabolic demand it is clearly important that aerobic and anaerobic metabolic pathways are tightly linked to energy requirements. There is a variety of ways in which the flux through a metabolic pathway can be controlled. Examples include regulation by substrate and product concentrations, and the modulation of individual enzyme activities by metabolic intermediates and cations such as calcium and

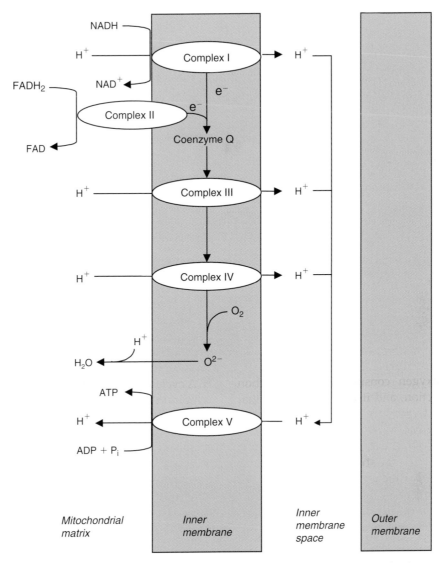

**Figure 6.5**  Outline of the electron transport chain occurring in the matrix and on (and between) the inner and outer membranes of the mitochondria. Electrons ($e^-$) derived from the reduced coenzymes NADH and $FADH_2$, enter the electron transport chain at complexes I or II, and the energy is used to transport protons ($H^+$) from the mitochondrial matrix to the inner membrane space. Passage of protons back through complex V provides the energy for ATP synthesis.

magnesium. In many cases enzymes can be phosphorylated or dephosphorylated, thus altering their affinity for substrates or moderators.

## Glycogenolysis

*Phosphorylase* is a key enzyme regulating the supply of substrate for glycolysis (see Fig. 6.3). The enzyme can be activated as a result of phosphorylation by *phosphorylase kinase*, which is itself regulated by another kinase whose activity is stimulated by cyclic AMP. Cyclic AMP is formed by the enzyme *adenylate cyclase* in response to adrenaline binding to receptors on the cell membrane, thereby providing for the hormonal control of phosphorylase (Fig. 6.6). Phosphorylase

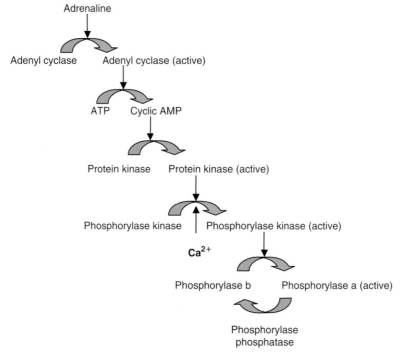

**Figure 6.6** Hormonal control of phosphorylase. The action of adrenaline initiates a cascade of phosphorylation reactions that lead to the activation of phosphorylase. Phosphorylase kinase is also activated.

kinase can also be activated by calcium at concentrations found in active muscles so that muscle contraction and glycogen breakdown are coordinated. A similar cascade of phosphorylation reactions controls glycogen synthase, but in this case they inactivate the process so that, when glycogen breakdown is stimulated during muscle contraction, synthesis is inhibited.

A *phosphatase* dephosphorylates phosphorylase so that activity is switched off rapidly when contraction ceases. Some of the products of metabolism, $P_i$ and AMP, also stimulate phosphorylase, but probably play a minor role in physiological regulation.

## Phosphofructokinase

Activation of *phosphorylase* may be thought of as a way of turning on the glycolytic pathway, whereas *phosphofructokinase* (PFK) is the main site at which glycolysis is controlled and restrained. Being a phosphorylation reaction, the PFK step is essentially irreversible. It is the rate-limiting step in glycolysis and is inhibited by a number of substances. Protons have a powerful inhibitory effect and high ATP concentrations also inhibit the enzyme by binding to a modulating site. Citrate, formed in the TCA cycle, is also inhibitory; this is interesting and important, because citrate tends to accumulate within the mitochondria when there is an energy balance (i.e. when the supply of ATP equals, or is greater than, the demand). Citrate diffuses into the cytoplasm where it inhibits PFK, thus forming a feedback mechanism with oxidative metabolism regulating glycolysis. When the muscle is under metabolic stress, ATP and citrate both decrease, thus releasing the inhibition on PFK.

PFK is also activated by raised levels of ADP and AMP. A further control mechanism has been described whereby fructose 2,6-biphosphate strongly activates PFK. Fructose 2,6-biphosphate (this is not the glycolytic intermediate fructose 1,6-biphosphate) is formed by the activity of a

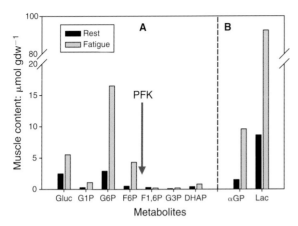

**Figure 6.7** Accumulation of glycolytic intermediates during a fatiguing isometric contraction. **A**, concentrations of glycolytic intermediates before and after the phosphofructokinase (PFK) reaction (arrow). **B**, accumulation of both α-glycerophosphate and lactate, which are needed to regenerate NAD from NADH. gdw, grams dry weight; Gluc, glucose; G1P, glucose 1-phosphate; G6P, glucose 6-phosphate; F6P, fructose 6-phosphate; F1,6P, fructose 1,6-biphosphate; G3P, glyceraldehyde 3-phosphate; DHAP, dihydroxyacetone phosphate; αGP, α-glycerophosphate; Lac, lactate. (Redrawn from the data of Edwards et al 1972.)

second form of PFK (PFK-2) that is itself activated by a cyclic AMP-dependent protein kinase. This constitutes the hormonal control of glycolysis in the liver, but it is not clear whether this mechanism is of importance in skeletal muscle.

The product of PFK, fructose 1,6-biphosphate, activates *pyruvate kinase*, the last enzyme of the glycolytic pathway. This is an example of feedforward control.

The importance of PFK as a rate-limiting step is seen by the fact that when a muscle is working hard and fatiguing there is a high glycolytic flux and an accumulation of glycolytic intermediates in the steps preceding PFK (Fig. 6.7).

## Intracellular pH and buffering

Under anaerobic conditions, when pyruvate and lactate accumulate, there is an acidification of the cell that has widespread consequences for the glycolytic pathway, for the function of contractile proteins and for ion transport mechanisms. If the quantity of lactic acid formed during a fatiguing contraction were in simple aqueous solution

(40 mM) the pH would be about 2, which would not be compatible with a living cell. However, muscle pH never falls below about 6.5, which indicates that the muscle fibre has a considerable buffer capacity. The main buffer in human skeletal muscle is the amino acid histidine, which is present in most proteins in the cell, followed by carnosine, with bicarbonate and inorganic phosphate making a small contribution. The role of inorganic phosphate is interesting because, although it is small at rest, it becomes a significant factor as phosphocreatine (PCr) is broken down and phosphate accumulates. The small increase in pH seen at the start of a contraction is due to PCr breakdown and $P_i$ accumulation increasing the buffer capacity. In contrast, the delayed recovery of pH seen after exercise is due to the fact that $P_i$ is decreasing as a result of PCr resynthesis. Mammals that depend on sustained high speed running for their survival, such as horses and dogs, have a higher muscle buffer capacity than humans, mainly due to higher levels of carnosine. Diving mammals such as seals and whales that work under anaerobic conditions for tens of minutes have exceptionally high levels of muscle carnosine.

Ultimately, intracellular pH is regulated either by the oxidation of pyruvate or by the co-transport of protons and lactate out of the muscle fibre. Transport of lactate out of muscle is a relatively slow process, and for this reason blood lactate levels remain high for 10 min or so after the end of high-intensity exercise. Lactate leaving a muscle fibre is taken up and metabolized by a variety of tissues including nearby inactive fibres in the same muscle, other resting muscles in the body, kidneys, heart and liver.

## Pyruvate dehydrogenase complex

Just as phosphorylase regulates the entry of carbohydrate into the glycolytic pathway, so PDH regulates the entry of pyruvate into the oxidative pathways. Like phosphorylase, PDH is subject to extensive controls including the phosphorylation and dephosphorylation of subunits, although in the case of PDH it is a calcium-stimulated *dephosphorylation* that activates the enzyme. ATP, NADH and acetyl-CoA all inhibit PDH activity by stimulating the regulating protein kinase. When fatty acids are oxidized, acetyl-CoA accumulates, inhibiting PDH and consequently reducing the utilization of carbohydrates.

## Regulation of the TCA cycle

In general, the activity of the TCA cycle is controlled by ATP levels. A high ATP concentration (and a low ADP level) inhibits PDH and thus limits the production of acetyl units; it also inhibits the electron transport chain, which is tightly coupled to the supply of ADP. Inhibition of the electron transport chain leads to an accumulation of reduced coenzymes and a shortage of NAD, required for the dehydrogenation reactions that drive the TCA cycle.

The TCA cycle can be thought of a wheel, or even a grindstone, that takes two-carbon fragments, breaks them down and disposes of

the remains as carbon dioxide and, eventually, water. Although the process is cyclical, activity can be limited by the concentration of some of the intermediates. For instance, oxaloacetate, which accepts the two-carbon fragment from acetyl-CoA, is at a low concentration in the mitochondria, and increasing the level stimulates TCA cycle activity. Another intermediate in the cycle, $\alpha$-ketoglutarate, is the acceptor for $NH_3$ in transamination reactions and consequently these reactions are a continual drain on TCA intermediates. The levels of TCA cycle intermediates are maintained by what are known as *anaplerotic reactions* (Fig. 6.8), using pyruvate or phosphoenolpyruvate as a substrate for carboxylation reactions producing oxaloacetate. An important process at the start of exercise is the transamination of glutamate catalysed by *glutamate–pyruvate transaminase*, which yields alanine and the TCA cycle intermediate $\alpha$-ketoglutarate. The need for a continuous supply of intermediates from

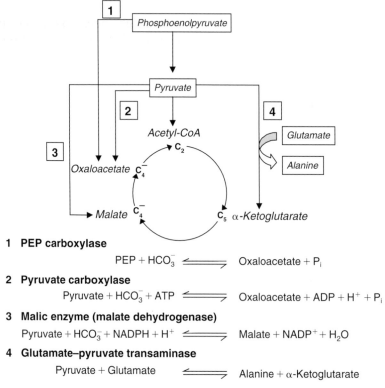

1 **PEP carboxylase**
$$PEP + HCO_3^- \rightleftharpoons Oxaloacetate + P_i$$

2 **Pyruvate carboxylase**
$$Pyruvate + HCO_3^- + ATP \rightleftharpoons Oxaloacetate + ADP + H^+ + P_i$$

3 **Malic enzyme (malate dehydrogenase)**
$$Pyruvate + HCO_3^- + NADPH + H^+ \rightleftharpoons Malate + NADP^+ + H_2O$$

4 **Glutamate–pyruvate transaminase**
$$Pyruvate + Glutamate \rightleftharpoons Alanine + \alpha\text{-Ketoglutarate}$$

**Figure 6.8** Anaplerotic pathways for TCA cycle intermediates. PEP, phosphoenolpyruvate.

carbon sources such as pyruvate is the reason why muscle requires some continuing carbohydrate metabolism during exercise, even when the major energy source is fatty acids.

## Electron transport chain

The electron transport chain is tightly coupled so that the flow of electrons down the chain is dependent on a supply of ADP, which is the main regulator of flux through this pathway. In addition, NADH or $FADH_2$ and $P_i$ are required together with oxygen, which is the final electron acceptor. Various poisons inhibit or uncouple the electron transport chain. Probably the most notorious is cyanide, which inhibits cytochrome oxidase, the last enzyme in the chain, preventing the use of oxygen as the final electron acceptor.

The vast majority of the ATP produced is used outside the mitochondrion and, to maintain a high flux, ATP inside the mitochondrion has to be exchanged for ADP on the outside. This process involves the adenine nucleotide translocase and mitochondrial creatine kinase (Fig. 6.9).

## MUSCLE COMPARTMENTS

Much of the discussion above, and indeed most literature on the subject, assumes that the cytoplasm of a muscle fibre is a homogeneous solution and that concentrations of metabolites and enzymes are uniform throughout the cell.

Thinking about internal membrane systems and the mass of myofibrils that pack the interior of a muscle fibre it is evident that there must be considerable barriers hindering diffusion. ATP is synthesized in the mitochondria, most of which are located near the cell membrane, but is utilized by myofibrils in the interior of the fibre, possibly 20 μm away. There will be a considerable concentration gradient between these sites, with ATP diffusing one way and ADP the other, but over this distance diffusion will be relatively slow. This problem may be partially overcome by the activity of creatine kinase (the phosphocreatine shuttle), where the high-energy phosphate is literally passed from one creatine kinase molecule to the next. The implication is that all that moves is the phosphate whilst the nucleotide portion of ATP or ADP remains relatively static (Fig. 6.9).

There are two forms of creatine kinase in muscles, one a cytosolic and the other a mitochondrial enzyme that is specified by mitochondrial DNA. The mitochondrial creatine kinase is bound to the inner membrane and is involved in translocating high-energy phosphate out of the organelle, keeping ADP levels high inside the mitochondrion and thereby driving the electron transport chain. Although creatine kinase is found throughout the cytosol, there are sites of higher concentration, specifically at places where ATP utilization is high, namely around the myofibrils and $Na^+/K^+$ pump, where the role of creatine kinase is to keep ATP levels high and ADP levels low.

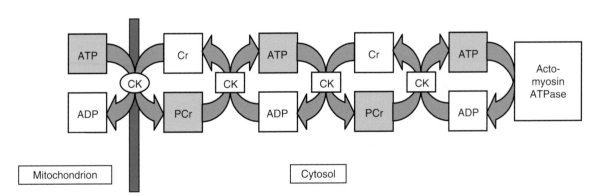

**Figure 6.9** The creatine kinase (CK)–phosphocreatine (PCr) shuttle. The scheme shows the transport of high-energy phosphate across the inner mitochondrial membrane (on the left) and its translocation to the myofibrils.

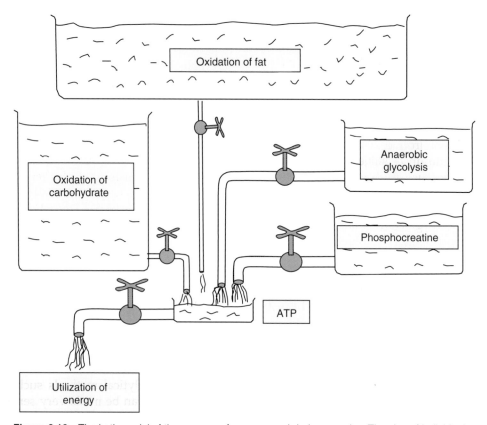

**Figure 6.10** The bath model of the sources of energy used during exercise. The size of individual baths indicates the approximate total energy available from the different substrates, except that those for oxidation of carbohydrate and fat are grossly underrepresented. Note that the diameter of the taps indicates the maximum rate at which ATP can be either used or resynthesized.

Similar forms of compartmentation exist for adenylate kinase and the glycolytic enzymes, and there may be local concentrations of glycolytic activity in parts of the cell where a supply of ATP is essential. Glycolytic enzymes are found associated with calcium release mechanisms in the T tubules, so there may be different ATP concentrations around the T tubules compared with the bulk of ATP in the cytosol.

## CONTRIBUTION OF DIFFERENT PATHWAYS TO ATP RESYNTHESIS DURING EXERCISE

The rate of ATP resynthesis varies between pathways and is inversely related to the total quantity of energy that is available (Fig. 6.10).

Very high-power outputs, such as are required during jumping or throwing, can be sustained for only a matter of seconds as a result of PCr breakdown. Power during sprinting (100–400 m) is achieved through a mixture of PCr breakdown and glycolysis (with the PCr probably used mainly to accelerate the body in the first phase of the sprint). The muscle content of ATP, PCr and the energy available from glycolysis is sufficient to sustain activity at a power output equivalent to 80% $V_{O_{2max}}$ for about 60 s. Oxidation of muscle glycogen stores can normally sustain the exercise for about 2 h. Liver glycogen and blood glucose are used primarily to prevent hypoglycaemia and to keep the brain functioning normally, but could sustain muscle activity at this level of effort for about another 20 min.

Because the rate of energy liberation from oxidation of carbohydrate is lower than that from glycolysis, the running speed during middle-distance events is about half that of sprinting. The level of muscle glycogen is very variable, depending on prior activity and diet in the 2–3 days before the exercise. Normal, well nourished, individuals have enough muscle glycogen to sustain exercise at around 70% $V_{O_{2max}}$ for about 2 h, although untrained subjects will find this very difficult because they will be limited by other factors such as pains in the legs and feet and rising body temperature. Although there are massive stores of triglyceride in the body, there is a limit to the rate at which this fuel can be utilized. Oxidation of body fat could, theoretically, support a high level of activity for hours, if not days, but in practice nobody can sustain this level after their muscle glycogen is exhausted. The maximum rate of energy production from the oxidation of fat alone is about 50% that from the oxidation of carbohydrate, so if exercise continues beyond the point where muscle glycogen stores are depleted, power output drops and running speed falls to little more than a fast walk. The lack of carbohydrate also leads to problems with depletion of TCA cycle intermediates, as discussed above.

# APPENDIX: TECHNIQUES FOR MEASURING MUSCLE METABOLITES

Cross-bridge function and calcium handling are very dependent on adequate energy reserves, and fatiguing exercise entails large metabolic fluxes and changes in the concentration of muscle metabolites. It is understandable, therefore, that a good deal of work in the past 30 years has been directed to finding an explanation for fatigue in terms of altered metabolite levels affecting force production by actin and myosin. For an investigation of force and metabolite changes, the first problem is to find a reliable method for measuring labile muscle metabolites.

Human muscle working in vivo can be sampled with a biopsy needle, plunging the needle plus sample of muscle rapidly into liquid nitrogen to freeze the tissue and stop biochemical reactions. Two or three samples can be taken from one muscle (usually large muscle groups such as the vastus lateralis, biceps and tibialis anterior) during the course of an experiment. At best, the time between taking the sample and freezing is around 5 s, an appreciable interval for metabolite changes to take place. Isolated muscle preparations can be frozen more rapidly, usually by flattening the muscle between two metal blocks cooled in liquid nitrogen, but, obviously, only one measurement can be made on each muscle.

For chemical methods, the tissue is frozen, extracted (usually with acid) and metabolites measured in the neutralized extract (de Haan et al 1986). Analytical methods such as linked enzyme assays can be made very sensitive and, with refinement, it is possible to make measurements on fragments of single fibres. More recently, high-performance liquid chromatography has been used and is very convenient for measuring the nucleotides, creatine and phosphocreatine, although lactate remains difficult by this method (Karatzaferi et al 1999).

Magnetic resonance spectroscopy (MRS) is increasingly used to measure phosphorus metabolites in muscle. In this technique, phosphorus nuclei are caused to process in a magnetic field and in so doing they radiate energy at radiofrequencies. The frequency of the radiation is characteristic of the chemical compound of which it is a part, and the amplitude of the signal is proportional to the quantity of the compound. Thus a Fourier analysis of the signal, which gives the energy at different frequencies, yields information about the composition and concentration of phosphorus metabolites present in the tissue. Developments in the technology of large superconducting magnets have led to the manufacture of MRS machines that can accommodate a whole

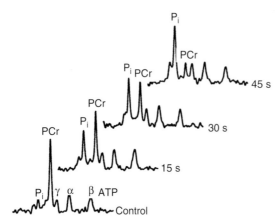

**Figure 6.11** Magnetic resonance spectroscopy used to measure phosphorus metabolites in human muscle (first dorsal interosseous). Averaged spectra from a fully rested muscle (control) and from a muscle after 15, 30 and 45 s of maximum voluntary contraction. Note the progressive decrease in phosphocreatine (PCr) and increase in inorganic phosphate ($P_i$) peaks. The $\alpha$, $\beta$ and $\gamma$ peaks of ATP remain constant.

body, but these instruments are used primarily for medical imaging. The great advantage of MRS is that measurements are non-invasive, and as there is no damage to the tissue the measurements can be repeated (Fig. 6.11).

The main disadvantage of MRS is that the radio signal is very weak and a number of spectra have to be averaged to obtain reliable data; for example,

about 1 min is currently required to obtain acceptable measurements from small muscles, whereas measurements from larger muscles can be made in 5–10 s. Consequently rapid changes in metabolites are difficult to follow. The apparatus is very large and the range of movement that is possible within such a machine is limited. The other disadvantage is that the measurements include a fixed volume of muscle and it is not possible to distinguish between changes in different fibre populations. One of the major differences between measurements made with MRS and those from biopsy specimens is the fate of ATP during fatiguing contractions. MRS measurements show little or no change, whereas biopsy measurements show appreciable changes. Part of the discrepancy may be due to larger changes occurring in the fast type 2 fibres.

The internal pH of the muscle fibre can be estimated from the position of the $P_i$ peak in the MRS spectra. As pH changes, so the charge on $P_i$ varies and the characteristic frequency will alter. Lactate cannot be measured directly from the phosphorus spectrum, but the change in lactate concentration can be calculated from the pH shift by making assumptions about the nature and quantity of intracellular buffers. The main buffers are protein-bound histidine, carnosine, bicarbonate and, in fatigued muscle, $P_i$.

REFERENCES AND FURTHER READING

Bessman S P, Geiger P J 1990 Transport of energy in muscle: the phosphoryl creatine shuttle. Science 211: 448–452
Brooks G A, Fahey T D, White T P 1996 Exercise physiology: human bioenergetics and its applications. California: Mayfield Publishing
De Haan A, de Jong J, van Doorn J E, Huijing P A, Woittiez R D, Westra H G 1986 Muscle economy of isometric contractions as a function of stimulation time and relative muscle length. Pflügers Archiv: European Journal of Physiology 407: 445–450
Edwards R H T, Harris R C, Hultman E, Kaijser L, Kohn D, Nordesjo L O 1972 Effect of temperature on muscle energy metabolism and endurance during successive isometric contraction, sustained to fatigue, of the quadriceps muscle in man. Journal of Physiology 220: 335–352

Karatzaferi C, de Haan A, Offringa C, Sargeant A J 1999 Improved high performance liquid chromatographic assay for the determination of 'high-energy' phosphates in mammalian skeletal muscle: application to a single-fibre study in man. Journal of Chromatography B 730: 183–191
Lowenstein J M 1990 The purine nucleotide cycle revised. International Journal of Sports Medicine 11: S37–S45
Matthews C K, Van Holde K E, Ahern K G 1999 Biochemistry. San Francisco: Addison Wesley Longman
Maughan R, Gleeson M G, Greenhaff P C 1997 Biochemistry of exercise and training. Oxford: Oxford University Press
Sahlin K, Katz A, Broberg S 1990 Tricarboxylic acid cycle intermediates in human muscle during prolonged exercise. American Journal of Physiology 259: C834–C841
Stryer L 1988 Biochemistry. New York: W H Freeman

1. What are the main pathways for the regeneration of ATP?

2. Under what circumstances are the different pathways used?

3. How are TCA cycle intermediates replenished?

4. How does the muscle cell deal with the fact that the sites of use and production of ATP are some disance apart?

5. How does the muscle cell deal with the fact that glycolysis requires the regeneration of $NAD^+$ during different intensities of exercise?

6. Describe the effects of high concentrations of ADP on fluxes through the different metabolic pathways.

7. What are the main regulatory sites for glycolysis?

8. What are the main regulatory sites for the aerobic metabolism of carbohydrate?

9. What evidence is there that PFK is a rate-limiting step in glycolysis?

10. What are the advantages and disadvantages of different analytical techniques for studying muscle metabolism?

# 7

# Muscle working in situ

In Chapter 2 the basic characteristics of force generation were described for an idealized muscle. This muscle would have tendons that did not stretch, it would be made up of parallel fibres, the sarcomere lengths would be constant along the length of the fibre, and the muscle would be fully activated. In practice, in the working body, almost none of these assumptions is true, and in this chapter we examine how these confounding factors affect the function of muscles working in situ.

## LENGTH–FORCE RELATIONSHIPS OF MUSCLE IN SITU

One of the major problems with the length–force relationship measured with anything other than short-fibre fragments is that, even if the whole muscle tendon complex is held at a fixed length (i.e. isometric), the active muscle fibres will shorten at the expense of series elastic elements, mostly the tendons. Consequently it is difficult to know the precise fibre length without measuring sarcomere or fibre lengths during the contraction, both of which are difficult to do with any precision. With intact human muscle it is impossible to obtain anything more than an estimate of the length of the muscle tendon complex, because it is difficult to specify the precise origin or insertion of a muscle, which may have fibres inserting at angles into a long aponeurosis. Instead of a length–force relationship, what is usually measured is the angle of the joint over which the muscle acts, giving an angle–force relationship.

There are two components of the total measured force of a contracting muscle (Fig. 7.1). The first is

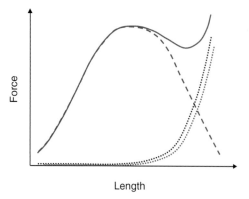

**Figure 7.1** Schematic length–force relations for total (solid line), active (dashed line) and passive (dotted line) force. The lower dotted line indicates the passive tension measured when shortening the muscle.

the active force due to the interaction of cross-bridges when the muscle is stimulated. The other component is the passive force that may result from a number of factors, connective tissue elements running in parallel with the muscle fibres or series elements within the muscle fibres, such as the titin filaments that hold the sarcomere together at long lengths (see Fig. 1.3). There may also be a small force due to the few cross-bridges that remain attached even in a resting muscle (short-range elastic component). To determine the active force generated at long muscle lengths it is necessary to subtract the passive force from the total. Although simple in principle, in practice this is always difficult because the passive force itself varies. Whatever gives rise to the passive force, it is not perfectly elastic and shows the property of hysteresis, that is, there is a greater passive force at the same overall muscle length when the muscle is being stretched than when it is shortening.

## FACTORS AFFECTING THE LENGTH–FORCE RELATIONSHIP

### Changes in fibre and sarcomere length during isometric contractions

The length–force relationship rarely shows the form illustrated in Figure 2.2A and there is often a significant discrepancy between the actual and

predicted values for the right-hand descending limb of the curve.

Close examination of the single-fibre preparations used in this type of work shows that the sarcomere spacing is not uniform along the length of a fibre, the sarcomeres at the ends of the fibre being shorter than those in the middle. If the muscle is set to a long length, so that all sarcomeres are on the right-hand arm of the force–length relationship, the sarcomeres at the ends of the fibre will be shorter and thus stronger than most. Consequently the force generated by the fibre will be more than that predicted from the average sarcomere length. The central sarcomeres will sustain this force because they can resist greater forces whilst being stretched than they can generate whilst isometric (see Figs 2.12 & 2.13).

## Level of activation

With increasing frequency of stimulation, the isometric force increases as the intracellular calcium level increases. As the length–force relationship is determined by the overlap of the thick and thin filaments, it would be expected that the force generated at all lengths would vary in the same way with changes in stimulation frequency. In practice, however, the shape of the length–force relationship is found to be dependent on the frequency of stimulation (Fig. 7.2).

As the frequency, and thus the force, increases, so the length at which maximum force is obtained moves to shorter muscle lengths. In the example shown in Figure 7.2, the optimum length decreases by about 3 mm, or 10% of the muscle length. Similar effects are also seen in skinned fibres (Stephenson & Williams 1982) activated with different concentrations of calcium and it seems that sarcomeres held at longer lengths are more sensitive to calcium.

## Shortening deactivation and lengthening activation

If a muscle, stimulated to develop isometric force, is allowed to shorten and then to redevelop tension, the final force is significantly lower than would be expected even taking into account the

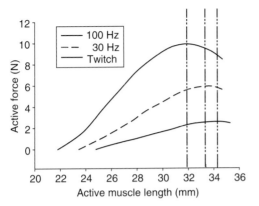

**Figure 7.2**  Length–force characteristics of rat medial gastrocnemius muscle when stimulated at three different frequencies: 1, 30 and 100 pulses per second. The vertical dotted lines indicate optimal length for the different stimulation frequencies. (Redrawn from the data of Roszek et al 1994.)

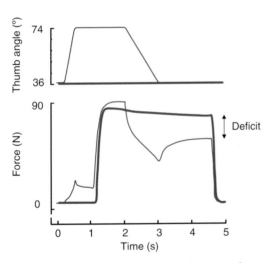

**Figure 7.3**  Superimposed force and angle traces of contractions of human adductor pollicis muscle with and without shortening, ending at the same angle between thumb and index finger. Thick traces are for the fully isometric contraction. The double-headed arrow labelled 'deficit' indicates the extent of shortening deactivation. (Redrawn from the data of de Ruiter et al 1998.)

shorter muscle length. This effect, known as 'shortening deactivation', has been seen in a wide variety of mammalian and amphibian muscle and single-fibre preparations. The effect is clearly demonstrated in the human adductor pollicis and was found to be proportional to the work done during shortening (Fig. 7.3). It was also

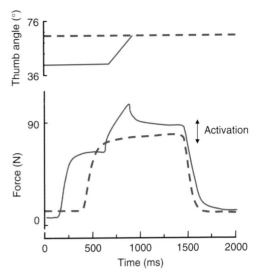

**Figure 7.4**  Superimposed force and angle traces of contractions of human adductor pollicis muscle with (solid line) and without (dashed line) a stretch, ending at the same angle between thumb and index finger. The double-headed arrow labelled 'activation' indicates the extent of stretch activation. (Redrawn from the data of de Ruiter et al 2000.)

largely reversed if the muscle was allowed to relax for as little as 200 ms. The mechanism of the effect is not known, but is suggested to be related to changes in sarcomere length along the muscle fibre occurring during the shortening step.

An opposite effect is seen when an active muscle is stretched and subsequently held at the new longer length. This effect, known as 'stretch activation', is illustrated for human adductor pollicis in Figure 7.4. Force enhancement following active stretch has also been described in a wide variety of preparations and, although the cause is unknown, is probably also related to unequal changes in sarcomere lengths along the fibre.

Resting muscle has a small but active force. A resting muscle will slowly shorten if it is not held at the ends and, if subjected to a small stretch, the muscle demonstrates elasticity (short-range elastic component; SREC). The elasticity is destroyed if the stretch is greater than about 1% of the fibre length. The SREC is generally thought to be due to a small number of cross-bridges that turn over very slowly. If stretched beyond their normal range, they detach and then reattach only slowly so that for a while the muscle loses its stiffness. This behaviour is

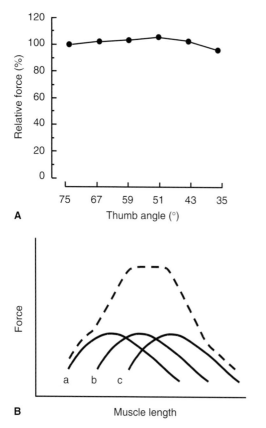

**Figure 7.5** **A**, length–force relationship of human adductor pollicis muscle. (Data from de Ruiter et al 1998.)
**B**, length–force relationship of a compound muscle group. The group is made up of three muscles, a, b and c, which have overlapping curves and combine to give a relationship shown as the dashed line.

sometimes referred to as the muscle being thixotropic. The effect can be seen in both isolated single-fibre preparations and whole human muscles in situ, and may play some role in maintaining muscle tone (Campbell & Lakie 1998, Hill 1968).

## Muscle morphology

Figure 7.5A illustrates the force generated by the human adductor pollicis over the full range of adduction; it is evident that there is little or no change in force at the extremes of length (or joint angle). One explanation for this might be that the fibres are very long and consequently are extended over only a small portion of their length–force curve, essentially staying on, or near, their optimum no matter what the position of the thumb. Dissection of cadavers and measurement of the length and position of the muscle fibre bundles shows that, on the contrary, the fibres are quite short and could easily be extended over their full range. However, the arrangement of fibre bundles within the muscle is such that they have overlapping length–force curves, so that with the thumb in full adduction there are some fibre bundles that are at their optimum length and when the thumb is moved to full abduction other fibres are then working close to their optimum (Fig. 7.5B). This arrangement of muscle fibres within a muscle means that a constant force can be exerted over a wide range of muscle lengths and probably applies to a greater or lesser extent to the majority of skeletal muscles.

## Angle of pennation

In a few muscles the fibres are arranged more or less in parallel; examples include the sheets of muscle found in the diaphragm and intercostal muscles. Elsewhere, the majority of skeletal muscles have a much more complex architecture. Monoarticular muscles are attached at the proximal end by a short tendon to bone and, distally, by a longer tendon passing over a joint. The muscle fibres insert into a sheet of tendon (aponeurosis) before this thickens to become the anatomical tendon. The size and shape of the aponeurosis varies between muscles. In some cases the aponeurosis runs along much of the muscle length, and the muscle fibres are relatively short and run at a large angle to the axis of the muscle. This angle is known as the angle of pennation. The direction of force generation of the muscle fibres is not the same as the direction of pull of the whole muscle, so the force acting along the muscle axis is the product of the muscle fibre force and the cosine of the angle of pennation. The angles of pennation normally found in mammalian muscles range from zero to around 30°. For an angle of 30° the resultant force acting along the axis of the muscle will be about 87% of the fibre force (cosine 30° = 0.87). This relatively small decrement in force is more than offset by an increase in the physiological cross-sectional area of the muscle compared with a parallel fibre muscle, as more fibres can be attached to

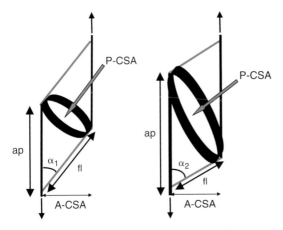

**Figure 7.6** Effect of pennation angle on muscle force. $\alpha_1$, $\alpha_2$ are angles of pennation; A-CSA and P-CSA are, respectively, the anatomical and physiological cross-sectional areas; fl, muscle fibre length from one aponeurosis to the other; ap, length of the aponeurosis. Note that for the larger angle of pennation the length of the aponeuroses increases, giving a greater area for attachment of large fibres; however, fibre length decreases so that A-CSA is the same in both cases.

the aponeurosis when they insert at larger angles of pennation.

Figure 7.6 shows two muscles of the same anatomical cross-sectional area (A-CSA) (i.e. the thickness of the muscle such as the biceps which might be measured with a tape measure), but with different angles of pennation. The muscle with the largest angle of pennation can pack more, or larger, fibres on to the aponeurosis; it consequently has a larger 'physiological' cross-sectional area (P-CSA) and can thus generate more force within the contractile parts of the muscle. There is a trade-off, however, between increasing pennation with more contractile material and the problem of resolving the force along the direction of the tendon, where the greater the angle the smaller the component of force (see above). The optimum pennation angle for generating force in the tendon is 45°. In the example given in Figure 7.6, the larger angle is greater than 45° and consequently, although the force developed in the contractile elements will be greater than is the case with a smaller angle of pennation, the force resolved along the tendon—the effective strength of the muscle—will be smaller. Figure 7.6 represents muscles at rest but when they contract, they change shape so the line of force runs from tendon

to tendon. This means that the angle between the aponeurosis and the tendon also has to be taken into account when resolving forces, together with any other deformations and changes in fibre length that may occur as tendons are stretched.

The differences in fibre length as a result of the change in angle of pennation, illustrated in Figure 7.6, will result in different length–force relationships of the two muscles. For the same change in overall muscle length, the fibres in the muscle with the larger pennation angle will be stretched or shortened relatively further and so have a smaller working range. (This also has implications for the force–velocity relationship.)

## Joints and levers

When working with human muscle it is not usually possible to measure the actual muscle length–force relationship, because it is impossible to attach a force transducer to the muscle itself. Instead what is normally measured is the relationship between joint angle and force measured at the wrist or ankle, and this introduces a new set of variables that have to be considered. Taking as an example the biceps brachii, it can be seen that the absolute force measured at the wrist will depend not only on the force generated by the muscle but also on the lever ratio. The critical ratio is that between the perpendicular distances from the point of rotation (the elbow) to the line of action of the forces ($a/b$), as shown in Figure 7.7. The consequences of a longer distance between insertion and point of rotation is that higher absolute forces will be generated but the working range will be reduced. There is a further complication. Figure 7.7C shows that as the elbow angle increases so the lever ratio decreases as the distance $a$ becomes vanishingly small. With the elbow flat the biceps will be pulling directly along the arm with virtually no vertical component of force. The way in which joint angle affects force varies widely, depending on the anatomy of the joint and the point of attachment of the muscle.

When measuring the isometric strength of a muscle it is essential that the position of the force transducer is standardized so that, for instance with the quadriceps, the strap is always

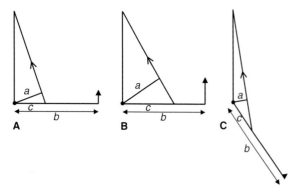

**Figure 7.7** Effects of joint angle on the lever arm and force measured at the wrist. *a*, perpendicular distance from point of rotation (elbow) to the line of action of the muscle force. *b*, moment arm—the distance from the point of rotation to the wrist where force is measured. *c*, distance from the point of rotation to the distal point of muscle attachment on the lower arm.

in the same position around the ankle. Devices that measure torque do not require such precise positioning of the ankle pad, but with these it is essential that the axis of rotation of the apparatus is aligned precisely with that of the joint.

# FACTORS AFFECTING THE FORCE AND POWER–VELOCITY RELATIONSHIP

## Role of tendons and other series elastic elements during movement

Until now in this chapter, the series elastic elements such as connective tissue and tendons have been treated as a nuisance, confusing the measurement of fibre properties, and a good deal of effort in muscle physiology has been devoted to ways of avoiding its influence. Nevertheless, in the living animal the elastic properties of the tendons play a major role in overall muscle activity. We have seen how, during an isometric contraction, the elastic tendons are stretched and this energy can be released during a subsequent shortening phase. The energy stored is greater if the shortening is preceded by a stretch of an active muscle, such as in a drop jump, and even during walking and running significant amounts

of energy are stored in the Achilles tendon. Many animals have this developed to a high degree. Horses and greyhounds have very long tendons in their legs allowing the rhythmic absorption and reutilization of energy as they run; this is taken to extremes by kangaroos, who literally bounce along (Alexander 1988). The tendon allows a high-power output or rapid movement to be generated by a slow muscle. Some insects, such as fleas, use muscles to bend their exoskeleton like a bow which, when released, gives them the power to jump far higher than could be achieved by the muscle alone. The human body cannot rival these feats but an interesting example is clicking the fingers. To do this, the thumb and middle fingers are opposed and a high-force isometric contraction developed, stretching the long flexor tendon of the middle finger. When, finally, the fingers slip, the middle finger moves far more rapidly than can be achieved by the muscle fibres alone and hits the skin hard enough to make a loud click. The speed of the movement is due to the rapid recoil of the long flexor tendon that had been slowly stretched by the muscle.

The force generated during a contraction in which there is shortening or lengthening may change because of a change in biomechanics of the joint over which it is acting or the fact that the muscle fibres move along the length–force relationship. Also, very few movements are smooth and at a constant velocity: almost always they start slowly with a relatively high force to begin the movement and then, as the limb accelerates, the velocity increases and the force that the muscle fibres can generate decreases. This decrease is, to some extent, offset by the presence of the tendons. During the initial high-force phase of the contraction, the tendons become stretched and then, as the movement begins, part of the overall shortening is taken up by the tendons. Thus the muscle fibres do not 'see' the full velocity of shortening that would be measured from the movement of the origin and insertion of the muscle. As the fibres are moving at a slower speed, the force they can generate during the movement is much higher than would be the case if there were no tendons.

The overall effect that tendons have on the force–velocity characteristics of muscle is to protect the fibres against high velocities. In the case of shortening, this increases the force and power that can be generated for any given velocity of shortening. When an active muscle is stretched, the effect of the tendons will be to reduce the overall force, as the fibres will be stretching more slowly than the whole muscle. As rapid stretching is potentially damaging to muscle fibres, the tendons play an important role in absorbing the impact of landing and protecting the fibres.

## Effects of angle of pennation and lever ratios on muscle speed and force

Up to a pennation angle of 45°, the isometric force of a muscle increases but, as can be seen in Figure 7.6, there is a corresponding decrease in the muscle fibre length and consequently the maximum velocity of shortening will be decreased. The net effect is to keep the power developed constant. Another effect of pennation is to change the relative velocity of fibre shortening. To generate a given degree of shortening in the whole muscle, individual muscle fibres in a pennate muscle have to shorten further than those in a parallel fibre muscle. Consequently, for any given velocity of whole muscle movement, the more pennate the muscle the closer individual fibres are working to their maximum velocity.

The effects of changing the lever arm (e.g. the biceps; Fig. 7.7) differ for the development of force, as measured at the wrist (i.e. the moment), and for the velocity of movement (the angular velocity). The closer the muscle inserts to the joint, the lower the force or moment measured at the wrist, but a given shortening of the muscle will result in a larger angular movement and a higher angular velocity. Power, being the product of force and velocity, will remain independent of the point of insertion, although the velocity at which maximum power is obtained will vary, just as it does when short fat and long thin muscles are compared (see Fig. 2.5).

## Level of activation

The velocity of shortening is generally thought to be independent of activation and cross-bridge number but, in practice, this is not the case. Work with fibre fragments suggests that for very small degrees of shortening the velocity is indeed independent of calcium, but for larger contractions the free calcium concentration is important (Moss 1986). It is longer contractions that are generally used when investigating muscle in situ, and Figure 7.8 shows how the shape of the force–velocity relationship, the maximum power and the velocity at which maximum power is obtained, all vary with the frequency of stimulation.

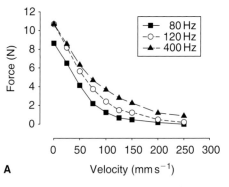
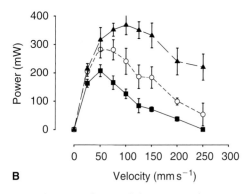

**A**   Velocity (mm s$^{-1}$)   **B**   Velocity (mm s$^{-1}$)

**Figure 7.8**   (**A**) Force–velocity and (**B**) power–velocity relationships of rat medial gastrocnemius muscle at different frequencies of stimulation. (Redrawn from the data of de Haan 1998.)

# APPENDIX: PROBLEMS OF MEASUREMENT

The force–velocity relationship is one of the fundamental properties used to define the contractile characteristics of a muscle, to differentiate fast and slow muscles, and to understand the process of fatigue. Unfortunately it is also one of the most difficult measurements to make with muscle working in situ. The main problems arise because of the interaction of the active muscle fibres with the series elastic elements in the whole muscle–tendon complex.

During a contraction, all elements in the muscle, the fibres, the aponeuroses and the tendons, experience the same force. The length of the series elastic elements (mainly the connective tissue and tendons) is proportional to this force. Consequently, if during a shortening contraction the force remains constant, the length of the series elements also remains constant and the whole of the shortening can be ascribed to the muscle fibres. This type of contraction is known as isotonic (same force) and was used in most of the early work on muscle function; a simple lever system with the muscle being used to lift a weight is all that is needed. However, there are problems with this type of system: one is the inertia of the mass to be lifted, another is that the velocity is not necessarily constant and a third difficulty is that, at light loads, the movement may begin before the muscle is fully activated. All of these problems can be overcome by mechanical adaptations to the simple lever system, but these become increasingly complex and difficult to apply to muscle working in situ in the human body.

Mechanical lever systems have been largely replaced by motor-driven systems but these are easiest to control when working at a constant velocity, which means that most measurements are now made in isokinetic (same velocity) mode. The difficulty here is that force is not controlled and may vary during the contraction. If the force is not constant, the length of the series elastic elements will also change and it is impossible to know for certain the velocity at which the muscle fibres are contracting.

Activation is a problem when working with both isotonic and isokinetic contractions, and is usually addressed by having the muscle contract isometrically until a steady maximal force is achieved before releasing the muscle tendon complex into the shortening phase. Figure 7.9 shows the type of relationship one would like to see. At the start of the release there is a phase in which the force falls rapidly, and during this time most of the shortening can be ascribed to the series elastic elements. Once the force falls to a constant level, the contraction is then essentially isotonic and it can be assumed that the muscle fibres are shortening at the velocity of the whole muscle–tendon complex. In practice, however, as can be seen in Figure 7.3, the force rarely reaches a constant value and some estimate has to be made of the velocity of muscle fibre shortening (e.g. Cook & McDonagh 1996, de Ruiter et al 1999). At low forces the tendons will be relatively stiff with respect to the muscle fibres and consequently most of the change in overall length of the muscle tendon complex is due to change in length of the muscle fibres. In this situation it is possible to estimate the velocity of the fibres from the velocity of the whole muscle. However, with higher forces, the muscle fibres become stiffer and more of the overall movement will be shared with the tendon, and it is then increasingly difficult

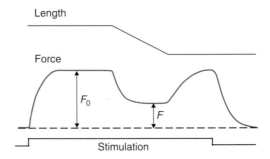

**Figure 7.9** Isokinetic shortening of an ideal muscle. The muscle is held at a fixed length while isometric force develops ($F_0$) and is then released at a constant velocity. Measurements of force ($F$) are made once a steady state has been reached during shortening.

to know the true velocity of muscle fibre shortening. There is a tendency, therefore, to overestimate muscle fibre velocity when dealing with high-force contractions.

A technique to avoid the initial rapid shortening phase is to use a pseudo-isotonic contraction. If the shortening starts at the time when the isometric force has built up to the level that it will be maintaining during steady shortening at the velocity being used, then the force is relatively constant for long enough to make a measurement (Fig. 7.10B). There are concerns about the extent of activation when using these short contractions, but higher frequencies of stimulation may offset this concern.

A complicating effect is the phenomenon of shortening deactivation described above, so that as the muscle shortens the force-generating capacity decreases. One of the origins of this decrease in force is an increase in sarcomere heterogeneity, as discussed above, but another feature occurring during shortening is that the muscle is moving along its length–force relationship. In the results from the adductor pollicis illustrated in Figure 7.3, this is not a serious problem as the relationship is relatively flat; however, in other muscles this may not be the case and there may be serious problems with any but very short movements.

One way of allowing for shortening deactivation is to extrapolate the force trace back to the start of shortening (Fig. 7.10A). This, however, is possible only for the lower velocities where the force trace is relatively linear (de Ruiter et al 1999). A similar technique can be used to deal with stretch activation. Starting with an isometric contraction, there is an initial rapid rise in force that is followed by a linear phase (see Fig. 7.4). The linear phase probably represents length-dependent enhancement of force due to increasing homogeneity of sarcomere lengths along the fibres. It is possible to extrapolate this phase back to the beginning of the stretch to obtain a value that is, hopefully, unaffected by series elasticity or length-dependent force enhancement.

The maximum velocity of shortening ($V_{max}$) can be estimated by extrapolating the velocities measured at lower loads using Hill's relationship, but is often uncertain because the curve approaches the velocity axis at a narrow angle and, as mentioned above, there is often difficulty in making measurements at high velocity. An alternative way of estimating $V_{max}$ in isolated preparations is the so-called 'slack' test. A contracting muscle is rapidly shortened by a known distance (at a velocity greater than $V_{max}$) so that the fibres become slack. The fibres will then shorten with no external load, and the time for

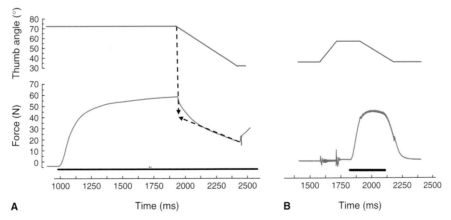

**Figure 7.10** Force and angle traces of shortening contractions of human adductor pollicis muscle. **A**, back extrapolation of the force to the beginning of shortening gives a value that is not affected by shortening deactivation. **B**, pseudo-isotonic contraction, where the shortening begins at the point where the isometric force reaches a value that can be sustained during the shortening phase. Some trial and error is involved in establishing the precise timing of the release. The thick dark line along the baseline indicates stimulation. (Redrawn from the data of de Ruiter et al 1998.)

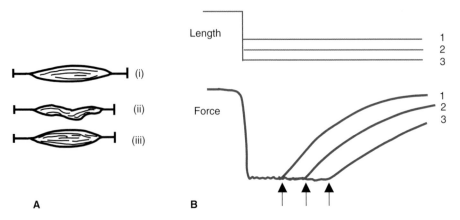

**Figure 7.11** Slack test to establish maximum velocity of shortening. **A**, (i) muscle contracting isometrically; (ii) muscle rapidly released; (iii) muscle shortens, takes up slack and redevelops force. **B**, muscle released for three different lengths. The time to begin redeveloping force (arrow) on the corresponding force record is used to calculate the velocity of unloaded shortening.

the fibre to take up the slack and begin pulling on the force transducer gives a measure of the velocity of unloaded shortening, sometimes known as $V_0$ (Fig. 7.11). In a mixed fibre muscle the extrapolated value of $V_{max}$ will be a complex function of all the fibres within the muscle, while $V_0$ is more likely to represent the maximum shortening velocity of the fastest fibres (Claflin & Faulkner 1985, but see de Haan et al 1989).

The work described above is applicable to small muscles such as in situ animal muscles and the human adductor pollicis, but there is a natural interest in large muscle groups such as the human quadriceps, the biceps and calf muscles. There are a number of commercially available isokinetic machines such as Cybex, which are widely used in sports science and physiotherapy for training,

rehabilitation and muscle function testing. The problems associated with measuring force–velocity relationships in small muscle groups are magnified when using large machines that generate high forces and moving large limbs. It is difficult to mix isometric and isokinetic modes in order to release an active muscle, the maximum speed of the movement is usually quite low, and problems with vibration and other artefact on the force trace are serious although, with care, limited measurements can be made (James et al 1994, 1995). Isokinetic testing of major muscle groups undoubtedly has a role to play in the routine assessment of athletes and patients but, in general, the results probably tells us more about the range of movement and problems in the knee than about the force–velocity characteristics of the muscle.

## REFERENCES AND FURTHER READING

Abbott B C, Aubert X M 1952 The force exerted by active striated muscle during and after change of length. Journal of Physiology 117: 77–86

Alexander R McN 1988 Elastic mechanisms in animal movements. Cambridge: Cambridge University Press

Campbell K S, Lakie M 1998 A cross-bridge mechanism can explain the thixotropic short-range elastic component of relaxed frog skeletal muscle. Journal of Physiology. 510: 941–962

Claflin D R, Faulkner J A 1985 Shortening velocity extrapolated to zero load and unloaded shortening

velocity of whole rat skeletal muscle. Journal of Physiology 359: 357–363

Cook C S, McDonagh M J 1996 Force responses to constant-velocity shortening of electrically stimulated human muscle–tendon complex. Journal of Applied Physiology 81: 384–392

De Haan A 1998 The influence of stimulation frequency on force velocity characteristics of in situ rat medial gastrocnemius muscle. Experimental Physiology 83: 77–84

De Haan A, Jones D A, Sargeant A J 1989 Changes in velocity of shortening, power output and relaxation rate during fatigue of rat medial gastrocnemius muscle. Pflügers Archiv 413: 422–428

De Ruiter C J, Sargeant A J, Jones D A, de Haan A 1998 Shortening induced force depression in human adductor pollicis muscle. Journal of Physiology 507: 583–591

De Ruiter C J, Jones D A, Sargeant A J, de Haan A 1999 The measurement of force–velocity relationships of fresh and fatigued human adductor pollicis muscle. European Journal of Applied Physiology 80: 386–393

De Ruiter C J, Didden W J M, Jones D A, de Haan A 2000 The force–velocity relationship of human adductor pollicis muscle during stretch and the effects of fatigue. Journal of Physiology 526: 671–681

Edman K A P, Elzinga G, Noble M I M 1978 Enhancement of mechanical performance by stretch during tetanic contractions of vertebrate skeletal muscle fibres. Journal of Physiology 281: 139–155

Edman K A P, Caputo C, Lou F 1993 Depression of tetanic force induced by loaded shortening of frog fibres. Journal of Physiology 466: 535–552

Gordon A M, Huxley A F, Julian F J 1966 The variation in isometric tension with sarcomere length in vertebrate muscle fibres. Journal of Physiology 184: 170–192

Hill D K 1968 Tension due to interaction between the sliding filaments in resting striated muscle. The effect of stimulation. Journal of Physiology 199: 637–684

James C, Sacco P, Jones D A 1994 An evaluation of different protocols for measuring the force–velocity relationship of the human quadriceps muscle. European Journal of Applied Physiology 68: 41–47

James C, Sacco P, Jones D A 1995 Loss of power during fatigue of human leg muscles. Journal of Physiology 484: 237–246

Moss R L 1986 Effects on shortening velocity of rabbit skeletal muscle due to variations in the level of thin-filament activation. Journal of Physiology 377: 487–505

Ramsey R W, Street S F 1940 The length tension diagram of isolated skeletal muscle fibres of the frog. Journal of Cellular and Comparative Physiology 15: 11–34

Roszek B, Baan G C, Huijing P A 1994 Decreasing stimulation frequency-dependent length–force characteristics of rat muscle. Journal of Applied Physiology 77: 2115–2124

Stephenson D G, Williams D A 1982 Effects of sarcomere length on the force–pCa relation in fast- and slow-twitch skinned muscle fibres from the rat. Journal of Physiology 333: 637–653

## QUESTIONS TO THINK ABOUT

1. Why does the length–force relationship of muscles at long length deviate from that predicted from the ideal length–force relationship based on the sliding filament model for sarcomeres?

2. What are the implications of sarcomere inhomogeneity for muscle fibres contracting at short lengths and at long lengths?

3. The way to define the optimal length of a muscle is to stimulate and record force at different lengths. Different answers are obtained when this is done using twitches compared with short tetani. Why is this?

4. How can stretch activation and shortening deactivation be demonstrated?

5. How do these phenomena affect attempts to measure the force–velocity relationships of a muscle?

6. How does the angle–force relationship of a muscle group such as the elbow flexors differ from the ideal length–force relationship of a single sarcomere? How do the elbow flexors compare with the quadriceps muscles in this respect?

7. If, as a result of training, the angle of pennation of the fibres became greater but the overall length of the muscle–tendon complex remained the same, what would happen to the isometric force of the muscle, the maximum velocity of shortening and the maximum power?

8. If the force remains constant during a shortening contraction of a muscle–tendon complex, what is happening to the length of the tendon and muscle fibres?

9. Draw the force–velocity relationship of a muscle tendon complex, (a) assuming the tendons to be very stiff and, (b) assuming them to be more compliant.

10. Is it possible to use isometric contractions to get an impression of the speed of a muscle?

# 8

# Effect of temperature and activity on muscle contractile properties

We saw in the previous chapter how factors such as series elasticity, fibre pennation and lever ratios all affect the force or velocity of a contracting muscle, or at least the movement of the limb to which the muscle is attached. These factors are relatively constant, the attachments and architecture of a muscle change little, and only slowly, over time, but we will now consider two other factors that can have a rapid and major effect on the contractile properties: temperature and previous activity.

## TEMPERATURE AND CONTRACTILE PROPERTIES

Mammals are warm-blooded and it might be thought that skeletal muscles would always be working at a constant temperature; this may be true of some muscles, such as the diaphragm situated deep within the body, but most are liable to considerable shifts in temperature. Small distal muscles such as those of the hands and feet may fluctuate from close to 40°C when working in a hot environment (washing up) to less than 20°C when outside on a cold day. Small muscles have both a small thermal capacity and are exposed to the elements, but even large muscle groups such as the quadriceps have temperature gradients of 3–5° from the outer to deep layers, and this is exaggerated if the subject is, for instance, standing in cold water. The majority of athletes warm up before exercise. While there may be a variety of reasons for doing this, one consequence is to increase the temperature of the muscle, either as a direct result of metabolic heat production or indirectly as a result of an increased blood flow bringing the muscle closer to core temperature.

Most experimental studies of human muscle in situ have investigated the effects of cooling rather than heating, partly because this may be a more common problem, certainly in northern climates. Even with isolated preparations there is not a lot of scope for increasing temperature, because above about 40°C isolated preparations rapidly deteriorate and die.

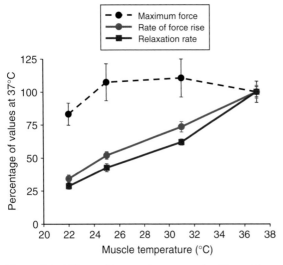

**Figure 8.1** Effects of temperature on maximal isometric force, maximal rate of force rise and maximal relaxation rate of human adductor pollicis muscle. (Redrawn from the data of de Ruiter et al 1999.)

## Effects of temperature on isometric force production

There is relatively little effect of temperature on the maximum isometric force generated by stimulation at high frequencies (50–80 Hz), there being only small changes in the range of 25–37°C (Fig. 8.1). The fact that maximum force production is relatively unaffected by temperature implies that the rates of cross-bridge attachment and detachment are affected to similar extents. The relative insensitivity of muscle force to temperature seen when stimulating at high frequencies is in contrast to the behaviour when stimulating at low frequencies (e.g. 10 Hz) (Fig. 8.2).

The reason for the difference between high- and low-frequency stimulation is evident from the records. Decreasing the temperature decreases the speed of the muscle with rates of both force development and relaxation decreasing (Figs 8.1 & 8.2). This has the effect of increasing the degree of fusion of the low-frequency tetanus, which shifts the force–frequency relationship to the left (Fig. 8.3). High frequencies of stimulation at low temperature results in a reduction in force, probably due to rapid failure of the electrical properties of the muscle, either the neuromuscular junction or conduction along the surface membrane or T tubules.

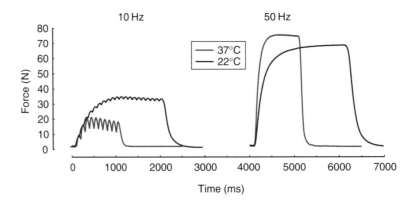

**Figure 8.2** Force records of electronically activated human adductor pollicis muscle at different temperatures and frequencies of stimulation. (Redrawn from the data of de Ruiter et al 1999.)

## Effects of temperature on velocity and power output

The changes in the rate of force development and relaxation with temperature suggest changes in cross-bridge kinetics, with major implications for force–velocity relationships and power output of the muscle. Figure 8.4 shows the force and power–velocity relationships for a range of temperatures that the adductor pollicis muscle in the hand may well encounter in normal life.

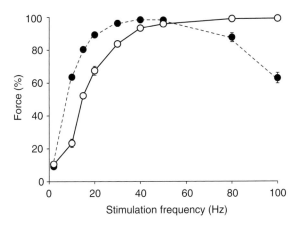

**Figure 8.3** Frequency–force relationships of human adductor pollicis muscle at 22°C (closed circles) and 37°C (open circles). Note that data are normalized for the maximal force in each condition. (Redrawn from the data of de Ruiter et al 1999.)

The major effect of low temperature is the much greater curvature of the force–velocity relationship and reduction in maximum velocity of shortening. The consequence of this change is a loss of maximum power, which is much greater than the loss of isometric force (Figs 8.4B & 8.5). The velocity at which peak power output is achieved is also shifted to the left in the colder, slower muscle.

## Warm-up and performance

Figure 8.4B shows the major effect that increasing muscle temperature has on power output. The benefits are dependent on the velocity of shortening, with the greatest effect seen at high velocities. In the case illustrated in Figure 8.4B, the improvement of power output at 50° per s on warming the muscle from 31 to 37°C was approximately 20%, whereas the improvement at 500° per s was over 100%. The expected benefit is, therefore, dependent on which side of the optimum velocity for power the muscle is working.

The optimum velocity for power output is very different for fast and slow fibres, the difference in $V_{max}$ being approximately fourfold. The common pedalling speed of 60–70 r.p.m. during cycling is probably a little above the optimum velocity for the slow type 1 fibres, and well below

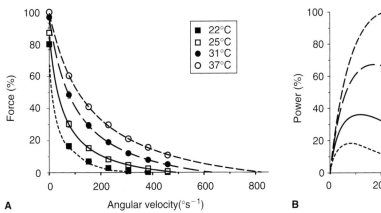

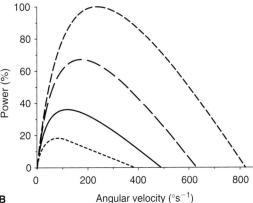

**Figure 8.4** Force–velocity (**A**) and power–velocity (**B**) relationships of human adductor pollicis at four different muscle temperatures. (Redrawn from the data of de Ruiter et al 2000.)

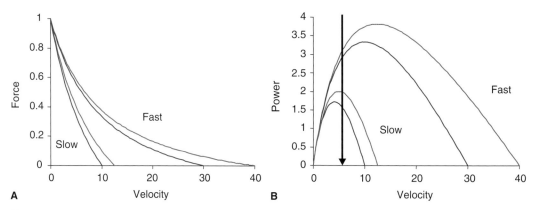

**Figure 8.5** Effects of temperature on performance of fast and slow muscles. **A**, force–velocity relationship; **B**, power–velocity curves. The effect of increasing $V_{max}$ by about 30% is shown. The arrow (**B**) indicates natural cadence during cycling. All values are in arbitrary units.

the optimum for the fast type 2 fibres (Fig. 8.5). Even competitive cyclists pedalling at a cadence of around 110–120 r.p.m. are probably still below the optimum for the fast fibres, while being far in excess of the optimum for the slow fibres (Sargeant 1994). Increasing the temperature has the effect of increasing the power output of both fibre types but, proportionately, the effect will be greater for the type 1 fibres at pedal frequencies above approximately 60 r.p.m. It can be seen in Figure 8.5B that at the speed indicated by the arrow there is very little improvement in power for the fast fibres as a result of warming the muscle while, for the slow fibres, there is a significant improvement. Even so, the effect of warming does not bring the slow fibres anywhere near the performance of the fast fibres.

These simulations produce the interesting prediction that the effects of warm-up on power output will be greatest for subjects with the largest proportion of type 1 fibres. Paradoxically the athletes most interested in power output (such as sprinters and throwers with a high type 2 fibre content) gain the least benefit from warming their muscles.

## ACTIVITY AND CHANGES IN CONTRACTILE PROPERTIES

Previous activity has significant effects on the contractile properties of muscle, and there may

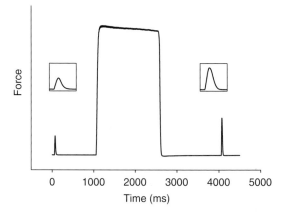

**Figure 8.6** Force traces of rat medial gastrocnemius muscle showing post-tetanic potentiation. Twitch contractions were given before and 1.5 s after a tetanus (duration 1.5 s, stimulation frequency 150 Hz).

also be changes during the course of a series of contractions used to test the muscle. Relatively short periods of exercise tend to enhance or potentiate muscle function and are discussed below. Longer periods of activity lead to fatigue and this is dealt with in later chapters (Chs 9 & 10).

## Post-tetanic potentiation

The size and shape of the twitch is affected by previous contractile activity, the classical example being known as post-tetanic potentiation (PTP) (Fig. 8.6). As a result of a single high-frequency

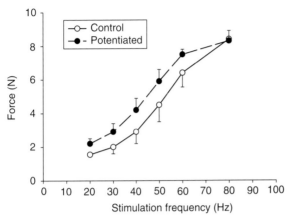

**Figure 8.7** Frequency–force relationships of rat medial gastrocnemius muscle for control contractions and shortly after a previous tetanus of 1-s duration with a stimulation frequency of 150 Hz (potentiated).

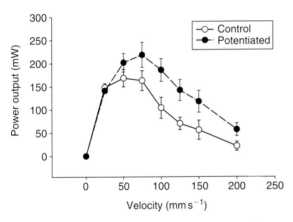

**Figure 8.8** Power–velocity relationships for rat medial gastrocnemius muscle for control contractions (stimulation frequency 80 Hz) and shortly after a previous tetanus of 1-s duration with a stimulation frequency of 150 Hz. (Redrawn from the data of Abbate et al 2000.)

contraction, twitch force can be more than doubled; the effect decays away fairly rapidly in the first 1–2 min, but may take as long as 10 min to disappear completely. Examining the shape of the twitch reveals that there are at least two changes: an increase in the peak force and a speeding up of both the rise of force and the relaxation phase. PTP is more evident in fast muscles: slow muscles tend to speed up as a result of a preceding

tetanus, but without increasing much, if at all, in peak force.

These changes in the twitch characteristics lead to changes in the force–frequency relationship for isometric contractions. The larger twitch means that there will be greater summation of force at low frequencies of stimulation but, at higher frequencies, when the muscle is more fully activated, the effect is minimal (Fig. 8.7).

A similar effect can be seen with the force and power–velocity relationships. As described in Chapter 7, a higher frequency of stimulation is needed to obtain maximum power from a muscle than to obtain maximum isometric force. Figure 8.8 shows that at 80 Hz where a previous tetanus has no effect on isometric force (Fig. 8.7), there is a significant increase in power, especially at the higher velocities. The effect is the result of an increase in the speed of the muscle (Abbate et al 2000).

## Catch-like properties

The barnacle has a smooth muscle that holds its two shells together; it has to maintain a high force for long periods and does so with relatively low energy expenditure. The muscle has a particular adaptation allowing it to develop this 'catch' state. Mammalian skeletal muscle cannot perform this trick, but in certain situations it can develop what is known as a 'catch-like' state. The term describes the increase in force that occurs when an initial, brief, high-frequency burst of stimuli (containing two to four high-frequency pulses) is included at the beginning of a subtetanic train of impulses (Burke et al 1970). Figure 8.9 shows the effect of an initial triplet on the isometric force of rat medial gastrocnemius muscle. Doublets or triplets at the start of a contraction also have the effect of increasing power output during shortening contractions at submaximal frequencies of stimulation. The catch-like effect is, in some ways, similar to that of PTP, but it is short lived, dying away after only a few tenths of a second. Although not extensively studied, the catch-like phenomenon seems to be greater with slow-twitch motor units and may offset some of the effects of fatigue (Binder-Mcleod 1995).

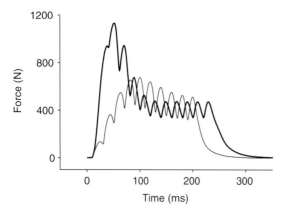

**Figure 8.9** Isometric force traces of rat medial gastrocnemius muscle showing the catch-like property. Stimulation frequency was 50 Hz (control; thin trace), or proceeded by a triplet (3 pulses at 400 Hz; thick trace).

## Consequences

The functional importance of PTP and the 'catch' phenomenon is not known. However, it is possible to imagine that these processes are an adaptation of the muscle to repeated use. If an animal has to make a series of maximal contractions, the power output of the second effort will be significantly greater. Alternatively, the same force or power output could be maintained with a lower neural drive—something that may be important during sustained activity. The catch-like effect tends to maximize the rate of force generation at the start of a contraction and there is good evidence of this happening with brief periods of high-frequency impulses at the start of voluntary contractions, especially during ballistic movements.

## Mechanisms of force augmentation

The mechanisms underlying PTP, staircase and catch are poorly understood. Clearly they may be explained by some combination of changes in calcium and cross-bridge kinetics that influence the size and shape of the twitch.

Cross-bridge function is altered by phosphorylation of the myosin light chains and it is likely that this will result in a faster rate of attachment and, possibly, movement through the power stroke (Sweeney & Stull 1990), giving rise to the larger, faster twitch seen with PTP.

Another, or additional, explanation for the increase in force in PTP and 'catch' is an increased calcium concentration in the myoplasm. The effect of a doublet or triplet will be to inject a burst of calcium into the fibre, with the level remaining raised for a brief time. This would explain the catch-like phenomenon. It is not known for certain whether calcium release is altered during PTP, but sites on the ryanodine receptor can be phosphorylated and dephosphorylated, and these may have a regulatory role. The sensitivity of the troponin binding sites can also be affected by phosphorylation of a regulatory site.

REFERENCES AND FURTHER READING

Abbate F, Sargeant A J, Verdijk P W L, de Haan A 2000 The effects of high frequency initial pulses and post tetanic potentiation on power output of skeletal muscle. Journal of Applied Physiology 88: 35–40

Binder-Mcleod S A 1995 Variable-frequency stimulation patterns for the optimization of force during muscle fatigue: muscle wisdom and the catch-like property. In: Gandevia S C, Enoka R M, McComas A J, Stuart D G Thomas C K (eds) Advances in experimental medicine and biology, vol 384: Fatigue; neural and muscular mechanisms, pp 227–240. New York: Plenum Press

Burke R E, Radomin P, Zajac F E 1970 Catch property in single mammalian motor units. Science 168: 122–124

De Ruiter C J, de Haan A 2000 Temperature effect on the force/velocity relationship of the fresh and fatigued human adductor pollicis muscle. Pflügers Archiv. European Journal of Physiology 440: 163–170

De Ruiter C J, Jones D A, Sargeant A J, de Haan A 1999 Temperature effect on the rates of isometric force development and relaxation in the fresh and fatigued human adductor pollicis muscle. Experimental Physiology 84: 1137–1150

Krarup C 1981 Enhancement and diminution of mechanical tension evoked by staircase and by tetanus in rat muscle. Journal of Physiology 311: 355–372

Ranatunga R W 1980 Influence of temperature on isometric tension development in mouse fast- and slow-twitch skeletal muscles. Experimental Neurology 70: 211–218

Sargeant A J 1994 Human power output and fatigue. International Journal of Sports Medicine 15: 116–121

Sweeney H L, Stull J T 1990 Alteration of cross-bridge kinetics by myosin light chain phosphorylation in rabbit skeletal muscle: implications for the regulation of actin–myosin interaction. Proceedings of the National Academy of Sciences USA 87: 414–418

## QUESTIONS TO THINK ABOUT

1. Which measurements of muscle function are the most sensitive to temperature?

2. With high-frequency stimulation temperature has little effect on the isometric force of a muscle but at low stimulation frequencies the colder muscle generates more force. Why is this?

3. At 22°C the maximum isometric force of the human adductor pollicis is reduced by about 20% compared with 37°C, but the maximum power is reduced by 80%. Why is there such a difference in these two measures of muscle function?

4. When cycling at a pedal frequency that gives maximum power output, what effect would warming the muscle have on the relative contribution of fast and slow motor units?

5. What are the consequences of post-tetanic potentiation for force production at high and low frequencies of stimulation?

6. What effect does post-tetanic stimulation have on the power that can be generated by a muscle? Why are the effects of post-tetanic potentiation greater at higher shortening velocities of the muscle?

7. How could you estimate maximum velocity of shortening in an isolated muscle or fibre preparation?

8. How could you estimate the effects of warming on the maximum velocity of shortening in a human muscle working in situ?

9. What would be the benefit of starting a contraction with a few fast repetitive pulses for in vivo movements?

10. Increased calcium release may be one cause of PTP. How could this be tested with an isolated muscle preparation?

# 9

# Introduction to fatigue

There are two main questions that are asked about fatigue. The first is which part of the complex contractile machinery has failed, and the second is why has it failed? This chapter deals primarily with the first of these questions, and the next chapter is concerned mainly with the second. Both this chapter and the next are concerned with high-intensity exercise, while Chapter 11 goes on to consider the special problems of endurance exercise.

## DEFINITION OF FATIGUE

The word fatigue has a number of different meanings: tired, exhausted, over-used, etc. Most of these words describe sensations that occur as a consequence of muscular activity of one sort or another, but it is also possible to experience fatigue without prolonged activity such as at the end of a difficult day that may have been mentally taxing but involved little physical effort. In addition, the same amount of physical work can feel very different if carried out in hot and humid conditions compared with a cool dry environment. It is important, therefore, when considering fatigue, to be clear about which particular aspect is under discussion: whether it is the sensations and perceptions of exertion or specific changes in muscle function.

Definitions of fatigue, even when limited to changes in muscle function, are complicated by the fact that the extent of fatigue may vary according to the method of testing. Changes of function may appear greater for voluntary contractions than for tetanic stimulation, or may differ according to whether the muscle is tested at one frequency of stimulation rather than another, or whether the contraction is isometric or shortening. It is important, therefore, in each situation to specify the type of change in muscle function that is being described as 'fatigue'.

A modest contraction can be sustained for some time without apparent failure or fatigue but if, at intervals during the contraction, the subject is asked to make a maximal effort, a progressive reduction in maximum performance will be seen. For this reason, fatigue is best defined as a loss of maximum or potential performance, being careful to define which aspect of performance is being considered—force, speed, power, etc.

## MANIFESTATION OF FATIGUE

Fatigue is one of the most common sensations of everyday life and, in a healthy person, can range from a pleasant feeling of tiredness to one of utter exhaustion after some herculean task. During short sprint events such as the 100 metres, the activity is all over before muscle performance begins to deteriorate seriously, but for 400 metres and over fatigue becomes an important factor. In sports such as football, hockey or tennis, fatigue is often of critical importance. Not only may running speed and power of the shot be affected, but it is when players become tired that skill begins to deteriorate and they make mistakes and concede goals or points.

Objective measurements of muscle function are difficult to make on the track or field, so most studies are laboratory based. It is also technically difficult to measure human muscle function whilst a person is moving, and consequently much of the work to date has used isometric contractions both to cause the fatigue and to monitor the changes in contractile properties.

Figure 9.1 shows a subject attempting to maintain a maximal voluntary isometric contraction for 1 min. The subject had feedback on his performance, he could see the force trace, and was loudly encouraged by his colleagues to maintain his effort. Without this feedback and encouragement, force declines very much more rapidly. In the last 30 s of exercise such as this, the pain in the muscle becomes severe and the subject has great difficulty in maintaining concentration.

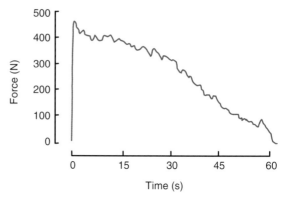

**Figure 9.1** Force loss during a sustained maximum voluntary contraction. The subject attempted to maintain a maximal contraction of the quadriceps muscle with visual feedback of the force record and with verbal encouragement.

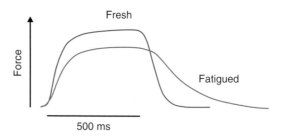

**Figure 9.2** A brief tetanus (50 Hz) before and after 45 s of fatiguing contraction. Human first dorsal interosseous of the hand; note the loss of force and the slowing of relaxation in the fatigued muscle. (Redrawn from the data of Cady et al 1989.)

Although useful measurements of contractile properties can be made from voluntary contractions by experienced subjects, there is always uncertainty as to how much of the fatigue is due to changes within the muscle and how much to a loss of voluntary control or motivation. The best way of examining changes in muscle function in an experimental situation is to test the muscle with brief electrically evoked contractions. Figure 9.2 shows a brief tetanus before and after 45 s of maximal voluntary contraction of a small hand muscle. There are two notable features of these records. The first is the loss of maximum force and the second is the slowing of the contractile response, most evident in the prolonged relaxation phase of the fatigued muscle. The question to be addressed is which aspect of the complex process is changing its function?

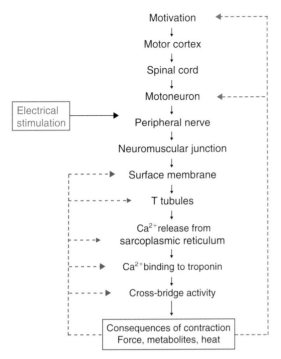

**Figure 9.3** Chain of command leading from motivation to the generation of force by the cross-bridges. The dashed lines indicate where the consequences of muscle activity might feed back and affect function either of the central nervous system (on the right) or of processes within the muscle itself (left). The large arrow indicates the point at which electrical stimulation can be used to distinguish between central and peripheral fatigue.

## THE CHAIN OF COMMAND

For muscle to function at all there is a series of events leading from the higher centres in the brain, via the spinal cord and motoneurons, and ending with the individual cross-bridges generating force (Fig. 9.3). A failure at any point in this chain will lead to a loss of force, and there are also several points at which a change in function could lead to a loss of force and slow relaxation.

## POSSIBLE SITES OF FAILURE

### Central factors during fatigue

It is quite possible that the activity of motor pathways in the central nervous system is modified during sustained activity.

Failure of central mechanisms can be assessed by superimposing electrical stimulation on voluntary contractions at various points along the chain of command (Fig. 9.3). This checks the integrity of the chain below the point of stimulation.

The motor cortex can be stimulated directly with high voltages or, more safely, with magnetic field stimulation. By carefully positioning the electrode over the motor cortex, specific muscles, or groups of muscles, can be activated. A similar technique can be used to stimulate motoneurons in the spinal cord. The information gained is largely limited to single twitches, but has been used to investigate changes in function of the motor cortex during sustained activity (Gandevia et al 1998).

A more widely applicable technique is to stimulate the peripheral motor nerves (Fig. 9.3). This can be achieved with a small button electrode placed over the anatomical nerve, such as the femoral nerve supplying the quadriceps or the ulnar or median nerves supplying muscles of the hand. An alternative is to activate the nerves through the skin (percutaneous stimulation). Large moistened pad electrodes are placed on the skin to stimulate the branches of the peripheral motor nerve as they enter the muscle. Stimulating the anatomical nerve has the advantage that the whole muscle can be activated, whereas with the percutaneous technique it is usually possible to stimulate only a quarter to a half of a large muscle. In this case the assumption has to be made that the portion stimulated is representative of the whole.

When stimulating the peripheral nerve it is possible to compare the force of an involuntary stimulated tetanus with that generated by a voluntary effort. The stimulated contraction does not depend on motivation or normal function of the central nervous system, and therefore any difference between the stimulated and voluntary contractions provides evidence of central, as opposed to peripheral, failure (Fig. 9.4).

Rather than stopping the voluntary contraction and then stimulating electrically, it is possible simply to superimpose electrical stimulation on top of the voluntary contraction. If there is any muscle not activated by the voluntary effort, this will be stimulated and will generate additional force. The more the muscle is activated by the voluntary effort, the smaller will be the additional stimulated component. Tetanic stimulation, especially if it is

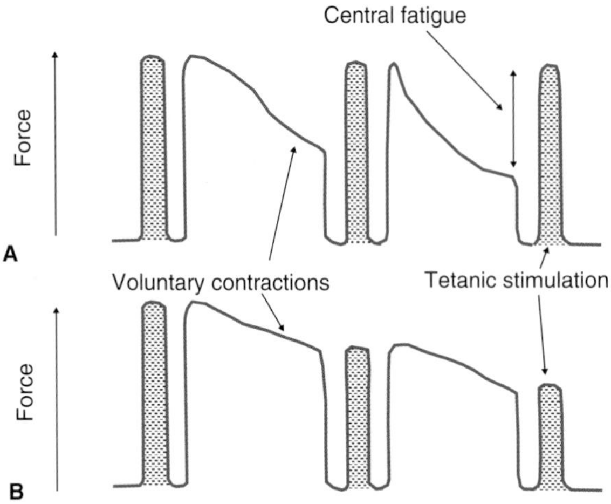

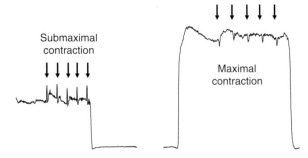

**Figure 9.4** Schematic figure showing the use of electrical stimulation to detect central fatigue. Voluntary contractions with interposed tetanic stimulation. **A**, fatigue partly due to central failure; note the greater fall in voluntary force compared with tetanic force. **B**, voluntary and tetanic force decline in parallel showing no central fatigue, i.e. the fatigue in this case is entirely of peripheral origin.

**Figure 9.5** Isometric contractions of the biceps with the muscle stimulated with single twitches (arrows), which the subject made either at the end of a submaximal contraction (**left**) or during maximal contraction (**right**). During the maximal voluntary contraction there was no additional force as a result of stimulation.

via a nerve supplying a large muscle group such as the quadriceps, can be painful and a good deal of work has used twitches rather than tetani (Rutherford et al 1986). If the muscle is stimulated with single pulses during a voluntary contraction, the size of the superimposed twitch will be inversely proportional to the degree to which the muscle is activated (Fig. 9.5). When stimulating during a voluntary contraction, the stimulated force characteristically goes up and then down below the level of force before the stimulus. This decrease is most evident during a maximal contraction, when the decrease is all that is seen. The drop in force is the result of the action potential blocking voluntary activation coming down the motor nerve for about 10 ms after the stimulus, so that there is a momentary involuntary relaxation of the muscle.

There has long been a controversy as to whether it is possible to activate skeletal muscle fully by a voluntary effort: it was believed that there was some inhibitory reflex, which protected the muscles or bones from possible damage. This myth was exposed when Merton (1954) showed that a maximum voluntary contraction of the adductor pollicis produced the same force as an electrically

stimulated contraction, and this observation has been confirmed for all the major muscle groups. Having said this, it is undoubtedly true that it is difficult to recruit an individual muscle fully; the perceived effort involved, the tremor and the failure if the subject if distracted, all emphasize the difficulty. In addition it is easy to think of actions and muscles over which the average person has little control, such as moving the scalp, ears and eyebrows, and one could include muscles used in belly dancing and a host of other exotic or erotic activities. Most of the work on recruitment and fatigue has made use of isometric contractions of single muscles or muscle groups. For some actions, such as knee extension with the quadriceps or elbow flexion with the biceps, this is a reasonably simple task, but even these simple actions are hardly 'natural'. Leaving aside research, it is uncommon to make prolonged maximal contractions of either biceps or quadriceps *in isolation*. Maximum contractions may occur many times when lifting and jumping, but this is always in the context of a motor programme involving other muscles and generally some movement. When it comes to small muscles, such as those in the hand or foot, most people have difficulty in moving them independently (e.g. the party trick of asking people to separate their fingers one at a time or, with even greater difficulty, their toes). Yet, when it comes to a 'natural' movement such as grasping and clenching a fist, it is almost certain that the individual muscles can then be fully activated.

Just as there have been doubts about whether it is possible to activate a fresh muscle fully, there has been similar controversy as to whether full activation can be maintained during fatiguing activity.

Using the type of approach illustrated in Figure 9.4, it has been shown that during sustained maximum isometric contractions of large muscle groups, such as the quadriceps, there begins to be a significant component of central fatigue after about 30 s, a time when the pain becomes difficult to ignore (Bigland-Ritchie et al 1978). With similar contractions of the adductor pollicis of the hand, however, there is little or no evidence of central failure. Gandevia et al (1998) have extensively examined the ability to activate the biceps; they found that most subjects can produce 95% of full activation both during isometric contractions and when the muscle is shortening, and that this does not change as the muscle fatigues. Similar conclusions have been reached for the quadriceps used during cycling (Beelen et al 1995) and repetitive knee extensions (James et al 1995) (Fig. 9.6). Making brief repeated maximal efforts appears to be easier than attempting to maintain a sustained isometric contraction.

## The neuromuscular junction

The idea that the neuromuscular junction (NMJ) may fail during fatigue is attractive, as exhausting the stores of neurotransmitter provides an obvious mechanism. At the start of a train of impulses the number of acetylcholine quanta released into the synaptic cleft falls but then comes to a steady level which depends on the frequency of release (i.e. the rate of stimulation) and the rate at which acetylcholine is resynthesized and made available for release. Where, as a result of poisoning (e.g. with curare or hemicholium) or disease (e.g. myasthenia gravis), the postsynaptic membrane has a reduced sensitivity or the number of quanta released is reduced, the fresh muscle may be close to the threshold for successful transmission. The first response to a train of impulses could be quite normal but, as the transmitter release decreases, the quantity falls below the threshold required to evoke an action potential in

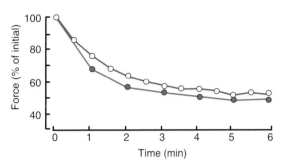

**Figure 9.6** Voluntary and stimulated contractions during repetitive knee extension. The subject made repeated knee extensions on an isokinetic ergometer; forces are plotted every 30 s (open circles). Every minute, the quadriceps was maximally stimulated via the femoral nerve (solid circles). Values are plotted as a percentage of the initial force. (Redrawn from the data of James et al 1995.)

the postsynaptic membrane and, consequently, force declines.

Working with an isolated rat phrenic nerve and diaphragm preparation, Krnjevic & Miledi (1958) concluded that at frequencies of around 10 Hz there was little likelihood of any failure, but that with higher frequencies breakdown might occur due to failure of conduction along the fine branches of the motor axons or of propagation along the surface membrane of the fibre. One way to test NMJ function is to compare the result of stimulating through the nerve with the force generated by direct stimulation of the muscle fibres. This procedure can be used with isolated nerve–muscle preparations, but it is difficult with human muscles in situ. Percutaneous stimulation with acceptable voltages activates the muscle only through the nerve endings, and direct activation of the muscle fibres requires very high voltages, which poses obvious difficulties for human subjects.

The effectiveness of electrical propagation across the NMJ in human subjects can be assessed more easily by stimulating the motor nerve and recording the muscle action potential. This signal is known as the *M wave* and is the summation of individual fibre action potentials in the part of the muscle near the electrode. Electrical activity is synchronized when a single electrical pulse is superimposed on a voluntary contraction, and a large M wave is recorded (Fig. 9.7). NMJ failure in a proportion of fibres will reduce the amplitude

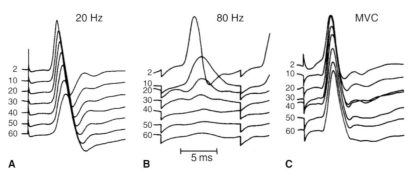

**Figure 9.7** Changes in the action potential waveform of the human adductor pollicis. Action potentials recorded every 10 s during a sustained isometric contraction. **A**, stimulation with a relatively low frequency of 20 Hz; **B**, with a high frequency of 80 Hz; **C**, single electrical shocks superimposed on a maximum voluntary contraction (MVC). (Redrawn from the data of Bigland-Ritchie et al 1979.)

of the M wave and the critical experiment is to see whether the M wave decreases in parallel with force as a result of fatigue. Merton (1954) was the first to use this technique to study fatigue in normal subjects and found that, while the force declined progressively during a sustained voluntary contraction, the muscle action potential remained constant, thus demonstrating the integrity of neuromuscular transmission. This conclusion was disputed by Stephens & Taylor (1972), who presented conflicting evidence, but since then Bigland-Ritchie et al (1982) have systematically examined the question of NMJ failure and were unable to find evidence of failure during maximal isometric contractions of up to 60 s. Figure 9.7 shows M waves recorded during tetanic stimulation and at intervals during a sustained maximal voluntary contraction. It can be seen that the M wave remains relatively constant in amplitude and waveform during the voluntary contraction.

## Electrical activity of the surface membrane

Once an action potential has been successfully generated at the NMJ, it then has to propagate along the length of the fibre and down the myriad T tubules into the interior of the fibre. Any failure in this process could lead to a loss of activation and a decrease in force, and there are experimental situations where this can be shown to occur.

During a prolonged tetanus the time course of force loss depends, in part, on the frequency of stimulation. At high frequencies (80–100 Hz in human muscles), force is maintained for only a few seconds before declining to about one-fifth of its initial value by 30 s. In contrast, when a muscle is stimulated at 20 Hz the force remains almost constant for 60 s or longer (Fig. 9.8A). After about 15 s of stimulation, more force is produced at 20 Hz, even though this frequency produced only 60–70% of the maximum force in the fresh muscle.

If, while stimulating at a high frequency, the frequency is decreased at a time when there has been a substantial loss of force, there is a rapid recovery of force (Fig. 9.8B). The recovery is unlikely to be due to any resynthesis of muscle metabolites as the time course is too rapid for this. In the examples illustrated, the muscle was stimulated via the motor nerve so the behaviour could be due to changes in the NMJ, but similar behaviour is seen when isolated, curarized muscle preparations are stimulated directly, suggesting that the failure is a feature of the muscle membranes and not of the NMJ (Jones et al 1979). This behaviour is sometimes known as 'high frequency' fatigue.

Recording M waves confirms that high-frequency stimulation leads to failure of electrical properties. Figure 9.7 shows that, although action potentials are well maintained when stimulating at 20 Hz, there is a rapid loss of amplitude and a prolongation of the waveform when stimulating

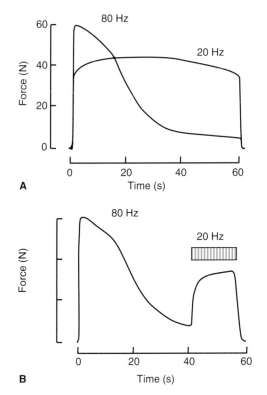

**Figure 9.8** Force loss at two frequencies of stimulation in human adductor pollicis. **A**, muscle stimulated continuously at 20 or 80 Hz. **B**, muscle stimulated at 80 Hz initially and then reduced to 20 Hz.

at 80 Hz. The reasons for this behaviour will be considered in the next chapter, but for the moment the question is whether this type of failure occurs during voluntary contractions and leads to a loss of force.

The M waves recorded during a maximum voluntary contraction (Fig. 9.7C) indicate that electrical failure is not a major problem during voluntary isometric contractions lasting for up to 60 s, as the amplitude and waveform are well preserved when compared to the rapid failure when stimulating at 80 Hz. There is, however, one reservation to this general conclusion. The signal recorded as an M wave is due largely to electrical activity in the surface membrane of the muscle fibre, whereas the activity in the T tubular membranes contributes relatively little. Consequently it is possible that conduction along the length of the fibre is well maintained during prolonged

contractions but there could still be problems with conduction along the T tubules. As will be discussed in Chapter 10, it is in these narrow tubules that $K^+$ is likely to accumulate and lead to problems with conduction.

## Excitation–contraction coupling and calcium release

Excitation–contraction (EC) coupling is the process whereby electrical activity in the T tubules leads to release of calcium stored in the sarcoplasmic reticulum (see Ch. 3). It is not possible to study this aspect of muscle function with human muscle in situ and the experimental work has to involve isolated preparations, often at the level of single fibres. Much of this work is discussed in the next chapter, but one relatively simple experiment, which suggests an involvement of EC coupling with fatigue, is presented here.

With isolated muscle preparations it is possible to circumvent the EC coupling mechanism and release calcium directly from the sarcoplasmic reticulum (SR) by using high concentrations of caffeine. Caffeine opens ryanodine receptors in the SR membrane, releasing calcium, which then activates the cross-bridges. If a normal force is produced, this suggests that cross-bridge function is preserved and that the fault lies in the process of EC coupling. The concentration of caffeine required to produce a contracture (so called because it is electrically silent) is around $20\,mmol\,L^{-1}$, orders of magnitude higher than could be obtained by drinking coffee, and this is one reason why the experiment has to be done with isolated muscle preparations.

The experiment illustrated in Figure 9.9 clearly shows that at a time when force is much reduced the loss can be reversed by artificially releasing calcium from the SR, the implication being that it is failure of the release process that is responsible for the fatigue.

## Cross-bridge function and slowing of the muscle

The work discussed in the previous section suggests that the main problem with fatigue is a failure

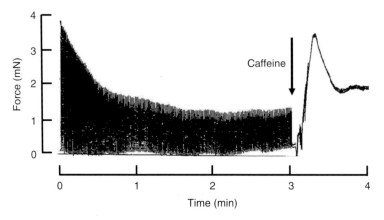

**Figure 9.9** Effect of caffeine on restoring force in a fatigued muscle. A small bundle of mouse fast muscle fibres was stimulated once a second until severely fatigued, when caffeine was added to the bathing medium. (Redrawn from the data of Jones & Sacco 1989.)

to release calcium from the SR and that cross-bridge function is preserved; in terms of one aspect of fatigue, the loss of isometric force, this may well be the case. However, a failure to release calcium does not explain the slow relaxation that is a feature of acutely fatigued muscle (see Fig. 9.2). Slow relaxation could be caused either by a slow re-accumulation of calcium into the SR or by a slowing of the rate of cross-bridge detachment.

The half-time of the exponential phase of the relaxation curve typically increases two- to three-fold as a result of fatiguing contractions in both human muscle working in situ and isolated animal muscle preparations. The relaxation phase is electrically silent, so the prolongation of force is not due to continuing activation of the muscle.

The relaxation rate during a series of repeated contractions, shown in Figure 9.10, changed little in the early part of a protocol but increased towards the end, giving an approximately threefold change over the 60 s of activity in the ischaemic hand muscle.

Contractile slowing and an increase in the duration of a twitch lead to an increased fusion of low-frequency tetani. Consequently there is a reduction in the frequency required to produce a fused tetanus and a shift of the force–frequency relationship to the left (Fig. 9.11). This is reminiscent of the difference between fast and slow muscles

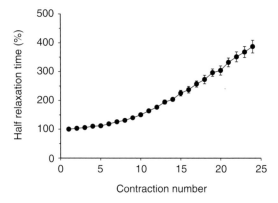

**Figure 9.10** Time course of slowing of relaxation during a fatiguing ischaemic protocol of human adductor pollicis muscle. (Redrawn from the data of de Ruiter et al 1999.)

(see Figs 4.6 & 4.7), so that essentially the effect of fatigue is (temporarily) to transform the muscle to a slower fibre type.

The slow relaxation from an isometric tetanus is accompanied by slowing of shortening velocity. Figure 9.12 shows force–velocity curves for a muscle in the fresh state and after being fatigued with a series of dynamic contractions. It can be seen that the estimated maximum velocity of shortening decreased by 33% with a similar decrease in isometric force. Similar findings can be seen with animal muscle preparations where it is also possible to verify the changes in maximum

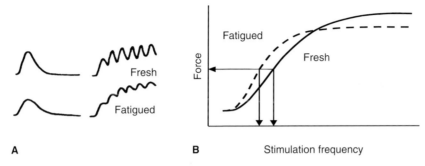

**Figure 9.11**  Effects of slow relaxation on the extent of fusion (twitch and stimulation at 7 Hz) (**A**) and the force–frequency relationship (**B**).

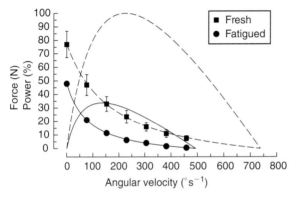

**Figure 9.12**  Force–velocity and power–velocity relationships for fresh and fatigued human adductor pollicis muscle. Dashed lines are for fresh muscle, solid lines for the fatigue condition. (Redrawn from the data of de Ruiter et al 2000.)

velocity of shortening directly by measuring unloaded shortening velocity with the slack test (de Haan et al 1989).

The shape of the force–velocity curve and the maximum velocity of shortening are properties of the cross-bridges and are not likely to be affected by changes in calcium release or reuptake into the SR. In Figure 9.12, the loss of isometric force could be due to a decrease in the amount of calcium released from the SR and/or to a change in the ability of cross-bridges to generate force. The maximum velocity of shortening, however, is a function of cross-bridge detachment at the end of the power stroke, and this provides direct evidence that cross-bridge function is altered with fatigue.

The major change in the force–velocity relationship is a loss of maximum power, which, proportionately, is much greater than the loss of isometric

force (Fig. 9.12). The velocity at which the peak power output is achieved is shifted to the left in the fatigued and slower muscle. This illustrates the point made in the definition of fatigue at the beginning of this chapter, that the extent of fatigue depends greatly on the mode of testing. Assessing fatigue with isometric contractions in this instance would indicate a modest loss of performance, whereas measuring power output at the optimum velocity shows a major decrement; the contrast would be even greater if measurements of power were made at higher velocities.

Different effects of fatigue under different measurement conditions are even more notable if shortening contractions are compared with the situation where an active muscle is stretched.

Figure 9.13 shows the force–velocity relationship for fresh and fatigued muscles extended into the region of muscle lengthening. Force is reduced throughout the whole range of movement (Fig. 9.13A) but, proportionately, the reduction is least when the muscle is stretched. If the forces are expressed as a percentage of the relevant isometric force then the force sustained by the fatigued muscle during stretch is greater than that of the fresh muscle (Fig. 9.13B). The results suggest that, although there may be fewer cross-bridges attached in the fatigued muscle compared to the fresh muscle, those that are attached are more resistant to stretch than they were when fresh.

## TEMPERATURE AND FATIGUE

Cooling a hand muscle from 37°C to 22°C makes little difference to the isometric force, and during

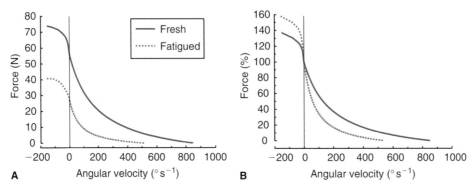

**Figure 9.13** Effect of fatigue on performance during eccentric contractions. **A**, absolute values for force; **B**, force normalized to the same isometric force. Human adductor pollicis fatigued with 30 shortening contractions under ischaemic conditions. (Redrawn from the data of de Ruiter et al 2000.)

a prolonged series of contractions there is little loss of force, slowing of relaxation, or change in shortening velocity or power output (de Ruiter et al 1999, 2000). The explanation is probably the effect of reduced temperature, which slows all metabolic processes in the muscle, and, superficially, it might appear that cooling a muscle would be a useful way of delaying fatigue. In reality, little would be gained from this if the muscle was needed to shorten or generate power, because reducing the temperature slows the contractile speed so much that in terms of power output it is effectively fatigued from the start. Cooling may be beneficial for muscles required for isometric contractions or if they are to resist stretching, but it is no benefit for power athletes.

## MOTOR UNIT FIRING FREQUENCY

During a maximum voluntary contraction of the human adductor pollicis, individual motoneurons discharge at different rates, ranging from about 10 to 50 Hz, and it has been shown that the mean frequency declines during fatiguing contractions (Bigland-Ritchie et al 1986). This decline is associated with slowing of the muscle relaxation rate (Fig. 9.14) and a shift to the left of the force–frequency relationship (see Fig. 9.11). These observations suggest that motoneuron firing rates for a fatigued muscle during a maximum voluntary contraction may be regulated to match changes in muscle contractile speed. A possible

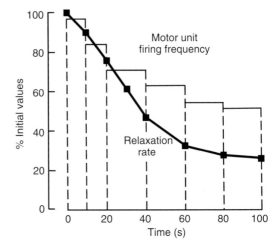

**Figure 9.14** Changes in mean motor unit firing rate recorded from fine wire electrodes in the muscle compared with changes in relaxation rate during a fatiguing maximum voluntary contraction of human muscle. Values are expressed as a percentage of the value in fresh muscle. (Redrawn from the data of Bigland-Ritchie et al 1986.)

advantage of this change in firing frequency is that it would avoid the problems associated with high-frequency stimulation (see Figs 9.7 & 9.8) where $K^+$ builds up in the extracellular spaces, reducing the excitability of the muscle.

There is no metabolic recovery if a muscle is kept ischaemic after a fatiguing contraction, and in this situation the motoneuron firing frequency remains low and the relaxation slow. This suggests there may be a reflex mechanism slowing motor unit

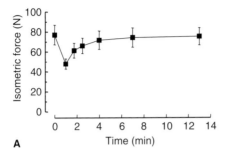

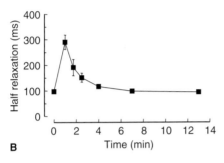

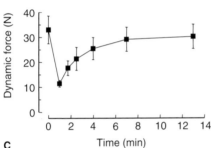

**Figure 9.15** Time course of recovery following a fatigue protocol applied to human adductor pollicis muscle in the first minute. **A**, maximum isometric tetanic force; **B**, half-time of relaxation; **C**, force at optimum velocity of shortening $(200° s^{-1})$ (i.e. power output).

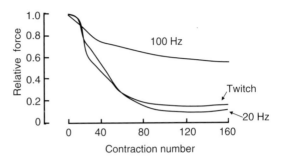

**Figure 9.16** Peak force developed by a twitch and 1-s tetani at 20 and 100 Hz tested at intervals during a long series of ischaemic contractions. Results are expressed as a fraction of the initial force, showing greater loss of force at low frequency than at high frequency of stimulation. (Data from Edwards et al 1977.)

## RECOVERY FROM FATIGUE

There are various phases of recovery following fatiguing exercise. Changes in the electrical properties of the muscle when stimulated at high frequencies recover rapidly within a matter of seconds. Maximum force, either voluntary or tetanic, together with all the indices of speed (relaxation rate, shortening velocity and power) recover with a half-time of about 1 min and are usually back to normal within 5–10 min (Fig. 9.15), although a small deficit of about 5% may persist for longer. This time course is consistent with that of metabolite recovery. Phosphocreatine and inorganic phosphate take about 2–3 min to recover, whereas intracellular pH takes somewhat longer.

### Low-frequency fatigue

Although most parameters of fatigue recover within minutes, there are other changes that may take much longer to recover. Such changes have been described in a number of different preparations and circumstances, and lead to a condition generally known as *long-lasting* or *low-frequency fatigue* (LFF) (Jones 1996).

After a long series of contractions there is generally a greater loss of force generated at lower, compared with higher, frequencies of stimulation (Fig. 9.16). This is a consequence of smaller twitches, although their general shape and speed is unaffected. The consequence of smaller twitches

firing in response to metabolic changes in the fatigued muscle (Bigland-Ritchie et al 1986). There are many receptors that might signal such changes. Group III and IV free nerve endings respond vigorously to many metabolites, such as $H^+$ and $K^+$, which accumulate in fatigued muscles. These nerve endings are implicated in cardiovascular responses to exercise and have powerful inputs to inhibitory motoneurons. In addition they reduce the discharge rates of γ motoneurons and therefore seem ideally suited to mediate a reflex that modulates motoneuron firing rates during fatigue.

is that less force is generated at lower stimulation frequencies and the force–frequency curve is shifted to the right. The shape of the force–frequency relationship is often described in terms of the ratio of forces developed at low compared with high frequencies of stimulation, 20 and 50 Hz often being used for human muscle. LLF produces a reduction in the 20 : 50 ratio. A notable feature of this form of fatigue is that the recovery is slow, taking hours and, on occasions, days, hence the alternative name *long-lasting fatigue*.

It has been a frequent observation that LFF is more pronounced in muscles that have been exercised eccentrically (i.e. stretched when activated) than during either isometric or shortening contractions. Thus, during stepping exercise, LFF is more pronounced in the quadriceps used to support the body while stepping down (eccentric contractions) despite the fact that these contractions are metabolically less demanding than the shortening (concentric) contractions of the opposite muscle used to step up. The slow recovery and the fact that it is more pronounced after damaging exercise suggests that LFF may be a consequence of damage or disruption to the EC coupling mechanism, so that less calcium is released per action potential.

## FIBRE POPULATIONS AND FATIGUE

There are major differences in the susceptibility to fatigue of the different fibre types. In general, the faster the contractile speed, the more rapid is the onset of fatigue. This can be demonstrated at all levels from single fibres to motor units and whole muscles, as well as being evident in the notorious inability of power athletes to compete over more than about 800 m. Figure 9.17 shows the fatigue profiles of mouse soleus (slow, mainly type 1 fibres) and the extensor digitorum longus muscle (fast, composed predominantly of type 2 fibres).

There are many reasons why faster fibres fatigue more rapidly than slow fibres. Because of their contractile speed, fast motor units require higher frequencies of activation and this could place a strain on the NMJ and EC coupling, although they may also have a higher density of $Na^+$, $K^+$ transport ATPase in the surface membrane to cope with the increased cation flux. Another is that the

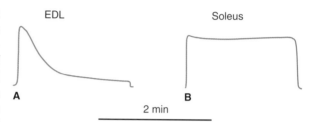

**Figure 9.17** Fatigue of fast and slow muscles. Isolated mouse muscles stimulated at 100 Hz for 0.25 s every second; 25°C. The tracings show the envelope of the force produced by each contraction. EDL, extensor digitorum longus.

fast fibres have a higher rate of ATP utilization than slow fibres because of the rapid cross-bridge cycling and the requirements of $Ca^{2+}$, $Na^+$ and $K^+$ pumping. Because energy supply for fast fibres has to be very rapid, only phosphocreatine and glycolysis are able to meet the demand, while the oxidation of carbohydrate and fat makes little contribution. Fast fibres have slightly more phosphocreatine, and appreciably more glycogen and glycolytic enzymes than slow fibres, but even these are sufficient for only a short burst of activity before there is an imbalance between supply and demand for energy.

## SUMMARY: SITES OF FAILURE

Over the past 50 years or more there has been a good deal of work and controversy, sometimes verging on the acrimonious, in trying to identify the weak point in the chain of command leading to muscle contraction. There have been strong advocates for almost every possibility: central failure, the NMJ, calcium release, cross-bridge action, etc. It is likely, however, that all are right to some degree, but all are wrong when they champion one site at the expense of all others. In designing a physical chain made of metal links there is no point in having one link weaker than the others. If one is made of thinner material, the additional material in all the other links is wasted—the other links are 'over-engineered'. The same principle applies to the physiological chain of events leading to muscle contraction. There is no point in having cross-bridges or calcium release mechanisms that can work under any condition if, say, the NMJ will fail after a few minutes of activity. This would be wasteful of protein and energy and, in evolutionary terms, would prove to be a disadvantage at a time when protein or energy was at a premium. It is likely, therefore, that evolutionary pressures ensure that most of the systems described above begin to fail at about the same time during stressful activity. Which actually fails first may depend on the particular type of activity or the experimental circumstances.

---

## REFERENCES AND FURTHER READING

Beelen A, Sargeant A J, Jones D A, de Ruiter C J 1995 Fatigue and recovery of voluntary and electrically elicited dynamic force in humans. Journal of Physiology 484: 227–235

Bigland-Ritchie B, Jones D A, Hosking G P, Edwards R H T 1978 Central and peripheral fatigue in sustained voluntary contractions of human quadriceps muscle. Clinical Science 54: 609–614

Bigland-Ritchie B, Jones D A, Woods J J 1979 Excitation frequency and muscle fatigue: electrical responses during voluntary and stimulated contractions. Experimental Neurology 64: 414–427

Bigland-Ritchie B, Kukulka C G, Lippold O C J, Woods J J 1982 The absence of neuromuscular function transmission failure in sustained maximal voluntary contractions. Journal of Physiology 330: 265–278

Bigland-Ritchie B, Dawson N J, Johansson R S, Lippold O C J 1986 Reflex origin for the slowing of motoneuron firing rates in fatigue of human voluntary contractions. Journal of Physiology 379: 451–459

Cady E B, Elshove H, Jones D A, Moll A 1989 The metabolic causes of slow relaxation in fatigued human skeletal muscle. Journal of Physiology 418: 327–337

De Haan A 1990 High-energy phosphates and fatigue during repeated dynamic contractions of rat muscle. Experimental Physiology 75: 851–854

De Haan A, de Jong J, van Doorn J E, Huijing P A, Woittiez R D, Westra H G 1986 Muscle economy of isometric contractions as a function of stimulation time and relative muscle length. Pflügers Archiv. European Journal of Physiology 407: 445–450

De Haan A, Jones D A, Sargeant A J 1989 Changes in velocity of shortening, power output and relaxation rate during fatigue of rat medial gastrocnemius muscle. Pflügers Archiv 413: 422–428

De Ruiter C J, de Haan A 2000 Temperature effect on the force/velocity relationship of the fresh and fatigued human adductor pollicis muscle. Pflügers Archiv. European Journal of Physiology 440: 163–170

De Ruiter C J, Jones D A, Sargeant A J, de Haan A 1999 Temperature effects on the rates of isometric force development and relaxation in the fresh and fatigued human adductor pollicis muscle. Experimental Physiology 84: 1137–1150

De Ruiter C J, Didden W J M, Jones D A, de Haan A 2000 The force–velocity relationship of human adductor pollicis muscle during stretch and the effects of fatigue. Journal of Physiology 526: 671–681

Edwards R H T, Hill D K, Jones D A, Merton P A 1977 Fatigue of long duration in human skeletal muscle after exercise. Journal of Physiology 272: 769–778

Gandevia S C, Herbert R D, Leeper J B 1998 Voluntary activation of human elbow flexor muscles during maximal concentric contractions. Journal of Physiology 512: 595–602

James C, Sacco P, Jones D A 1995 Loss of power during fatigue of human leg muscles. Journal of Physiology 484: 237–246

Jones D A 1996 High and low frequency fatigue revisited. Acta Physiologica Scandinavica 156: 265–270

Jones D A, Sacco P 1989 Failure of activation as the cause of fatigue in isolated mouse skeletal muscle. Journal of Physiology 410: 75P

Jones D A, Bigland-Ritchie B, Edwards R H T 1979 Excitation frequency and muscle fatigue: mechanical responses during voluntary and stimulated contractions. Experimental Neurology 64: 401–413

Krnjevic K, Miledi R 1958 Failure of propagation in rats. Journal of Physiology 140: 440–461

Merton P A 1954 Voluntary strength and fatigue. Journal of Physiology 123: 553–564

Newham D J, Jones D A, Clarkson P M 1987 Repeated high force eccentric exercise: effects on muscle pain and damage. Journal of Applied Physiology 63: 1381–1386

Newham D J, Jones D A, Turner D L, McIntyre D 1995 The metabolic costs of different types of contractile activity of the human adductor pollicis muscle. Journal of Physiology 488: 815–820

Rohmert W 1960 Ermittlung von Erholungpausen für statische Arbeid des Menschen. Internazionale Zeitschrift für Angewante Physiologie 18: 123–164

Rutherford O M, Jones D A, Newham D J 1986 Clinical and experimental application of the twitch superimposition technique for the study of human muscle activation. Journal of Neurology, Neurosurgery and Psychiatry 49: 1288–1291

Sjogaard G, McComas A J 1995 Role of interstitial potassium. Advances in Experimental Medicine and Biology 384: 69–80

Stephens J A, Taylor A 1972 Fatigue of maintained voluntary muscle contraction in man. Journal of Physiology 220: 1–18

---

## QUESTIONS TO THINK ABOUT

**1.** Muscle fatigue is difficult to define. Which fatigue parameters should, in your opinion, be taken into account in a definition?

2. Why are effects of fatigue more pronounced when the muscle is tested at a relatively high shortening velocity compared with a low velocity?

3. Which are the possible links in the chain of command where changes could lead to slow relaxation?

4. Is the eccentric force sustained by a fatigued muscle higher than that in fresh muscle? If so, why?

5. Firing frequency decreases during a sustained voluntary isometric contraction. What would happen if there was no accompanying slowing of relaxation?

6. What processes might determine the rate of muscle recovery following fatiguing exercise?

7. What happens to potassium and sodium concentrations inside and around a muscle when it is stimulated at high frequencies?

8. Which elements in the chain of command could give rise to the loss of amplitude and slowing of the M wave seen when stimulating at high frequencies?

9. Why do fast and slow muscles fatigue at different rates? Why is it not possible to have a fast muscle that does not fatigue?

10. What evidence is there that cross-bridge function changes as a result of fatigue arising from high-intensity exercise?

# 10

# Mechanisms of fatigue

Although it was concluded in Chapter 9 that all elements in the chain of command could fail in the right, or wrong, circumstances, this chapter will be concerned mainly with three links of the chain that seem to be particularly affected by high-intensity exercise lasting for 60 s or so. These are the electrical properties of the surface membrane of the fibre, the processes of calcium release from the sarcoplasmic reticulum and the function of cross-bridges. Central fatigue has a role to play in the latter stages of this type of activity, but in this chapter we are concerned only with the mechanisms of peripheral fatigue.

## ELECTRICAL PROPERTIES OF THE MUSCLE FIBRE

We have seen that muscle function is very sensitive to high-frequency stimulation, and the loss of force is accompanied by a reduction in the amplitude and slowing of the time course of the M wave (see Fig. 9.7).

During high-frequency stimulation there are changes in the concentrations of sodium and potassium in and around the muscle fibres, which can lead to loss of excitability and decrease in force. With every action potential there is an influx of sodium into the muscle fibre and an efflux of potassium. The potassium flux from a working muscle is large and may be sufficient to increase the extracellular concentration from the normal value of around $4 \, mmol \, L^{-1}$ to as high as $9 \, mmol \, L^{-1}$ (Hnik et al 1976, Sjögaard et al 1985). The T tubular membrane has a very large surface

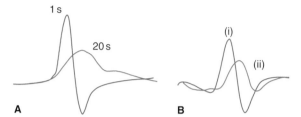

**Figure 10.1** Changes in the action potential waveform. **A**, human adductor pollicis; action potentials recorded after 1- and 20-s stimulation at 50 Hz. **B**, isolated mouse diaphragm; action potential of fresh muscle in medium containing (i) 4 mmol L$^{-1}$ K$^+$ and (ii) 15 mmol L$^{-1}$ K$^+$.

area across which potassium ions pass into the small volume of the T tubules themselves. Diffusion along the T tubules to the surface of the fibre is relatively slow so that potassium concentrations can build up rapidly to high levels within the depths of the T tubular network.

Increasing the potassium ion concentration in the extracellular medium of an isolated muscle causes a rapid loss of force and slowing of the action potential conduction velocity (Fig. 10.1), mimicking changes seen in human muscle fatigued by high-frequency stimulation (compare with Fig. 9.7).

These observations suggest that the muscle fibre membrane becomes depolarized during high-frequency stimulation (Jones & Bigland-Ritchie 1986) and direct measurements of membrane potential in single fibres confirm that the potential falls from −80 mV to −40 mV in about 10 s when stimulated at 70 Hz (Lännergren & Westerblad 1986, 1987).

It is not immediately obvious why partially depolarizing a muscle fibre should lead to failure of the action potential and loss of force as, in theory, it should be easier to initiate an action potential in a partially depolarized muscle. In practice, partial depolarization of a muscle fibre leads to a slow inactivation of sodium channels and this blocks the generation of an action potential, or at least reduces the amplitude and the speed of propagation along the surface membrane and down the T tubules (Jones 1996). It is for this reason that high extracellular potassium levels are potentially very dangerous, because potassium has a similar

action on the heart and a concentration of 8 or 9 mmol L$^{-1}$ would be fatal if injected into a subject at rest. This raises the interesting question of why athletes do not suffer from heart attacks, even though their plasma potassium concentration may rise to apparently dangerous levels during exercise. The answer lies in the Na$^+$, K$^+$ transport ATPase, which transports potassium into the cell and sodium out. The pump is electrogenic in that it transports 2K$^+$ inwards for 3Na$^+$ out, thereby making the interior of the cell more negative with respect to the outside. This hyperpolarization counteracts the effect of the extracellular accumulation of potassium, making the muscle membrane more difficult to depolarize. The same is true for the heart (Sjögaard & McComas, 1995). The transport ATPase is regulated by a wide range of factors, but catecholamines released during exercise are very potent in stimulating activity, as is the accumulation of sodium within the muscle fibre (Nielsen & Clausen 2000). The reduction in motoneuron firing frequency during prolonged contractions (see Motor unit firing frequency in Ch. 9) helps to reduce the cation fluxes and minimize the problem. Increased blood flow in working muscles also helps by washing away K$^+$ accumulating in the extracellular spaces, and this potassium will be taken up by inactive muscles and other tissues by their membrane cation pumps, which have been stimulated by circulating catecholamines.

## CHANGES IN CALCIUM RELEASE

If the neuromuscular junction and surface membrane are working correctly, the next stage to consider in the chain is the process of calcium release. The experiment illustrated in Figure 9.9 suggests that there is a problem with calcium release in the fatigued muscle fibre and there have been a number of similar observations (Eberstein & Sandow 1963, Westerblad & Allen 1991). The experimental approach in this field has been greatly strengthened as a result of the development of calcium indicators that can be injected into fibres allowing changes in intracellular calcium concentration to be measured.

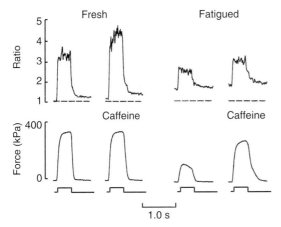

**Figure 10.2** Effects of caffeine (10 mmol L$^{-1}$) on contractile force of mouse single muscle fibres. Lower panel, force records with stimulation periods shown below the records. Upper panel, calcium signal is expressed as the ratio of fura-2 fluorescence at two excitation wavelengths. Force is given in kilopascals (force per unit cross-sectional area of the fibre). (From Westerblad & Allen 1991, with permission.)

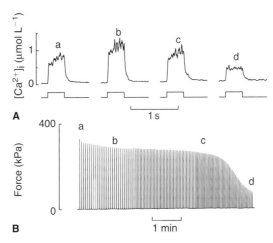

**Figure 10.3** Records of intracellular Ca$^{2+}$ levels (**A**) and force (**B**) obtained from a single mouse muscle fibre during a fatigued run. In A, the records are from the times indicated in lower-case letters above the force record. Periods of stimulation are shown below the calcium records. (From Westerblad & Allen 1993b, with permission.)

At low concentrations (1–5 mmol L$^{-1}$), caffeine potentiates the release of calcium via the ryanodine receptors, so that with each action potential there is a larger release of calcium and a larger twitch. With higher concentrations (15–20 mmol L$^{-1}$), caffeine opens the ryanodine receptor without the T tubular membrane having to be depolarized, and calcium floods out to cause a contracture.

Figure 10.2 shows the effect of a sub-contracture concentration of caffeine on a fresh muscle where, despite the fact that additional calcium is released, the force of a maximum tetanus is unaffected, indicating that the troponin is fully saturated with calcium in the fresh state. The first thing to notice in the fatigued state is that the calcium signal is reduced and, second, that the effect of caffeine is again to increase the calcium concentration but, this time, the force is also increased significantly. This shows that in fatigued muscle the troponin is not fully saturated with calcium, primarily because of a reduced calcium concentration in the sarcoplasm. However, this does not rule out the possibility of a reduction in the affinity of troponin for calcium, which could be affected by a range of factors including pH and inorganic phosphate ions.

## Calcium release

Figure 10.3 shows records of force and intracellular calcium during a fatiguing protocol. In this protocol the recovery interval between the tetani was reduced in three stages. In the first two stages there was a modest loss of force (about 15%) but this was accompanied by no change, or even a slight increase, in the amount or calcium released. During the third phase, however, when there was a major loss of force, it can be seen that there was a concomitant reduction in intracellular calcium levels (Westerblad & Allen 1993b), indicating that during this last phase it was a failure of calcium release that was the major reason for force loss.

## Calcium binding to troponin

Calcium binding to troponin is a critical stage in force generation and could well be affected in fatigued muscle. Observations on skinned fibres have shown that decreasing pH and increasing inorganic phosphate concentration shifts the calcium–force relationship down and to the right, with fast fibres being rather more sensitive than slow fibres (Donaldson & Hermansen 1978). The relationship between intracellular calcium and

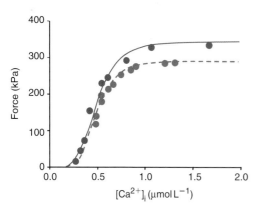

**Figure 10.4** Relationship between steady-state intracellular $Ca^{2+}$ levels and force of single mouse muscle fibres in control (solid line) and fatigue (dashed line) conditions. (From Westerblad & Allen 1993, with permission.)

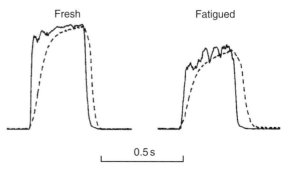

**Figure 10.5** $Ca^{2+}$-derived (solid lines) and real (dashed lines) force compared in fresh and fatigued mouse muscle fibre. $Ca^{2+}$-derived force was obtained by converting intracellular $Ca^{2+}$concentration records into force by means of the steady-state $[Ca^{2+}]_i$–force relationship. Note that, although there is a slow relaxation of real force in the fatigued muscle, the $Ca^{2+}$-derived force does not show the same slowing. (From Westerblad & Allen 1993b, with permission.)

force with intact mammalian fibres (Fig. 10.4) shows a decrease in force at saturating levels of calcium, which reflects a decrease in the ability of cross-bridges to develop force. There is also a small shift to the right in the relationship, so that a slightly higher concentration of calcium is required to produce half maximal activation in the fatigued muscle compared with fresh (Westerblad & Allen 1993b). However, the effect is not large and seems unlikely to be a major cause of force loss, at least in mammalian muscle.

## Calcium removal

Slowing of relaxation from an isometric contraction is a characteristic feature of fatigued muscle and there is a long history of speculation that this is due to a slow re-accumulation of calcium by the sarcoplasmic reticulum. Direct measurements of intracellular calcium levels mean that this idea can now be tested. In amphibian muscle fibres slowing of relaxation is, indeed, associated with slow calcium reuptake in the sarcoplasmic reticulum (Lee et al 1991). However, in mouse muscle fibres the observation is that slowing of force relaxation is not caused by slower calcium re-accumulation (Fig. 10.5) (Westerblad & Allen 1993b, Westerblad et al 1997).

## CROSS-BRIDGE FUNCTION AND FATIGUE

There are several indications that cross-bridge function is impaired during fatigue. The data in Figure 10.3 show that in the first phase of the fatiguing protocol there was a decline in force that quite clearly was not a consequence of a reduced calcium concentration and must represent (a modest) failure of the cross-bridges. Likewise, the data in Figure 10.4 show a relatively small loss of force at saturating levels of calcium, and we may conclude from this that failure of cross-bridge function leads to a drop of up to 20% in isometric force in fatigued muscle. Slowing of contractile function (both slowing of relaxation from an isometric tetanus and a reduction in maximum velocity of shortening) is a prominent feature of fatigued muscle (see Ch. 9). Change in the maximum velocity of shortening can be accounted for only by a change in cross-bridge kinetics and is therefore a key indicator of altered cross-bridge function in fatigue.

## CHANGES IN MUSCLE METABOLITES

Changes in calcium release and in cross-bridge function are associated with high-intensity activity

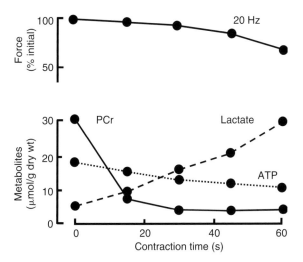

**Figure 10.6** Muscle metabolites during a sustained contraction. Isolated mouse soleus muscle (25°C) poisoned with cyanide (to ensure no aerobic recovery) and stimulated at 20 Hz. PCr, phosphocreatine. (Redrawn from the data of Edwards et al 1975.)

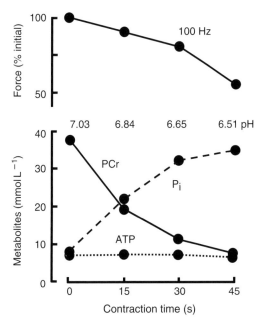

**Figure 10.7** Muscle metabolites during a sustained contraction. Human first dorsal interosseous was fatigued by a 15-s maximum voluntary contraction and then tested by stimulating at 100 Hz. MRS spectra were collected over the next 3 min. The cycle was repeated three times with the muscle maintained ischaemic throughout. Numbers above the metabolite values are the intracellular pH. PCR, phosphocreatine; $P_i$, inorganic phosphate. (Redrawn from the data of Cady et al 1989a.)

where energy supply is unable to match demand. The changes in function are exacerbated if the muscle is metabolically compromised, working in anaerobic conditions or in the presence of metabolic poisons such as cyanide or iodoacetate. Moreover, muscle function is restored only if metabolic recovery occurs and the time courses of function and metabolic recovery are very similar. Consequently, when looking for reasons why calcium release or cross-bridge function are compromised in fatigue, it is natural to see how muscle metabolites may influence these processes. To this end we will first review the metabolic changes that occur in a fatiguing muscle.

The changes in muscle metabolites during a sustained maximal isometric contraction are shown for an isolated mouse soleus muscle (Fig. 10.6) and the human first dorsal interosseous, studied in situ (Fig. 10.7).

For the mouse muscle, metabolites were measured by chemical methods and for the human muscle by magnetic resonance spectroscopy (MRS). The change in force during the exercise is also shown. Despite the differences in species and analytical methods, the results for the two preparations are basically similar. Phosphocreatine (PCr) concentration falls rapidly during the early stages

and there is a concomitant rise in inorganic phosphate ($P_i$) levels. Lactate concentration increases in a linear fashion throughout the contraction. The lactate was measured directly in the mouse muscles but with the human muscle lactate levels were calculated from measurements of intracellular pH, making assumptions about buffer capacity.

In Figure 10.6, lactate began to accumulate from the start of the contraction indicating that glycolysis was activated from the start of the contraction, so there is no sign of the often-quoted 'alactic' phase. Reports of a delay have arisen because the rise in blood lactate is delayed at the start of exercise because of the time it takes for muscle lactate to appear in the circulation.

In the human experiments (Fig. 10.7) the ATP, measured by MRS in the whole muscle remained relatively constant, while in the mouse the ATP declined. The different findings for the two preparations may be partly methodological.

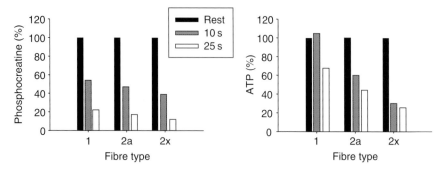

**Figure 10.8**   Measurements of metabolites in different fibre populations of human quadriceps before (rest) and 10 and 25 s after maximal cycling exercise. Phosphocreatine and ATP values are normalized for the values measured in resting fibres in each subject. (Redrawn from the data of Karatzaferi et al 2001a,b.)

In trying to assess the importance of different metabolites in the generation of muscle fatigue it is often useful to block different pathways with metabolic poisons such as iodoacetate, which inhibits glycolysis, or cyanide, which inhibits mitochondrial respiration. There are also a number of rare metabolic diseases with specific enzyme deficiencies that afford an important insight into muscle function. One that has been studied extensively is myophosphorylase deficiency or McArdle's disease. These patients lack phosphorylase, the first enzyme of the glycolytic pathway; they cannot break down glycogen and therefore produce no lactate during exercise. Consequently, instead of their muscles becoming acidotic during exercise, they become slightly alkalotic as a result of the breakdown of PCr (Cady et al 1989a).

Measurements of metabolites made on whole muscle cannot detect differential changes in the various fibre types. Dissection of individual fibres from human biopsy samples shows that there are differences in the rates of metabolite changes during short maximal exercise on a cycle ergometer (Fig. 10.8) (Karatzaferi et al 2001b).

The large decreases in ATP concentration seen in the fastest fibres are mirrored by increases in inosine monophosphate (IMP) formed as a result of the action of myokinase and AMP deaminase removing the ADP released by ATP hydrolysis. Some 90% of ATP is in the form of MgATP. When this is hydrolysed and converted to IMP, most of the $Mg^{2+}$ is liberated, leading to an increase

in the free $Mg^{2+}$ concentration from approximately $0.5\,mmol\,L^{-1}$ at rest to $2.5\,mmol\,L^{-1}$ after intense exercise. Free $Mg^{2+}$ may affect calcium release (see below).

During repetitive activity where the circulation is intact, there is an opportunity for the muscle to partially recover between each contraction so that, with appropriate work–rest cycles, a steady state is reached. Figure 10.9 shows the extent of metabolite changes during repetitive activity with light and heavy loads in human muscle measured with MRS.

The changes in ATP, PCr, $P_i$ and lactate levels described so far are, quantitatively, the major differences that occur with fatigue. However, it must be remembered that there will also be many other changes. For instance, there will be small increases in ADP, AMP, $NH_3$ and alanine levels, as well as many glycolytic and tricarboxylic acid cycle intermediates. Muscle glycogen concentration will also decrease, and in the fastest fibres this loss may be considerable (Kernell et al 1995). In the following discussion of mechanisms, however, only the major metabolic changes are considered.

## INFLUENCE OF METABOLITES ON FORCE PRODUCTION

Various experimental approaches have been used to determine the effect of altered metabolite levels on muscle function. One way is to look for correlations between metabolite changes and function,

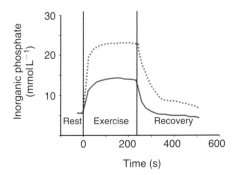

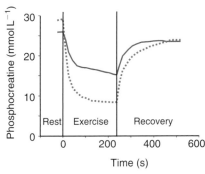

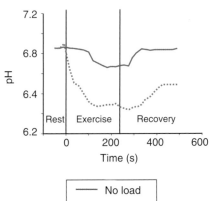

——— No load

········· Extra load

**Figure 10.9** Muscle metabolites measured by MRS at rest, during repetitive exercise of the lower leg, and in recovery. Repetitive knee flexion was performed inside the magnet with and without an extra load. (Redrawn from the data of Yoshida & Watari 1992.)

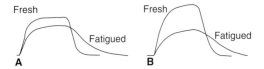

**Figure 10.10** Fatigue with and without $H^+$. **A**, normal subject, test tetani before and after 45-s fatiguing contraction. **B**, patients with McArdle's disease tested before and after 21-s fatiguing activity. (Redrawn from the data of Cady et al 1989a.)

using a detergent, which makes the surface membrane permeable. The composition of the external incubation medium can then be altered and the effect on force or shortening observed.

## Influence of metabolites on calcium handling

The calcium release mechanism is very complex with numerous possibilities for regulation and failure (see Ch. 3), and measurements of intracellular calcium levels clearly show that calcium release is reduced under conditions of metabolic stress, whether induced by contractile activity or metabolic poisons (Westerblad et al 1998). The accumulation of lactate and $H^+$ in fatiguing muscle is frequently cited as the cause of fatigue, and it is undoubtedly true that decreasing pH can have serious effects on a range of enzyme activities. There are, however, good reasons for doubting that this is the whole story. The first reason is based on observations of patients with McArdle's disease. If $H^+$ were the cause of the problem, these patients would not fatigue, but their clinical problem is that they fatigue much more rapidly than normal subjects. The records in Figure 10.10 clearly show that McArdle's patients suffer from the same changes in contractile function as normal subjects—loss of force and slowing of relaxation—despite not accumulating $H^+$ during the fatiguing activity. Consequently it is unlikely that $H^+$ is the cause of changes in calcium release, and this point has been directly demonstrated by Westerblad & Allen (1993a), who have shown that acidification increases calcium release. Curiously, making fibres alkalotic has the opposite effect (Fig. 10.11).

As $H^+$ is not the cause of decreased calcium release, attention has turned to other metabolites;

as in Figures 10.6 & 10.7, but another important technique is to use skinned preparations—either single fibres or small bundles. Single fibres can be skinned mechanically (with great care) by removing the surface membrane with a needle, but the more common method is chemical skinning,

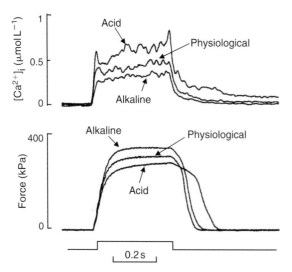

**Figure 10.11** Effect of altered intracellular pH on calcium release and force. Single mouse fibres incubated in a bicarbonate buffer gassed with 5% carbon dioxide to give a physiological pH, and with 30% and 0% carbon dioxide to give a 0.5 pH-unit intracellular acidification or alkalinization, respectively. Note the opposite effect of pH on the calcium and force traces. (Redrawn from the data of Westerblad & Allen 1993a.)

current interest centres on changes in ATP, $Mg^{2+}$ and $P_i$, either separately or in combination.

## Magnesium

Magnesium has an inhibitory effect on calcium release. The resting intracellular $Mg^{2+}$ concentration is about 1 mmol $L^{-1}$ and reducing this to 15 $\mu$mol $L^{-1}$ induces spontaneous calcium release in skinned fibre preparations (Lamb & Stephenson 1994). Conversely, increasing the $Mg^{2+}$ concentration to 10 mmol $L^{-1}$ abolishes depolarization-induced calcium release. However, when $Mg^{2+}$ was injected into resting fibres, the decrease in force was not so marked (Westerblad & Allen 1992b).

## ATP

Although calcium release is blocked at very low ATP concentrations, Lamb & Stephenson (1991) could demonstrate very little effect in the range of 2–8 mmol $L^{-1}$ ATP in skinned fibre preparations. Subsequently it has been suggested that high

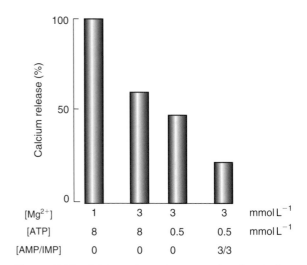

**Figure 10.12** Calcium release from sarcoplasmic reticulum in the presence of various concentrations of magnesium and muscle metabolites. Values are expressed as a percentage of release in 'fresh' conditions ($Mg^{2+}$ 1 mmol $L^{-1}$, ATP 8 mmol $L^{-1}$). (Redrawn from the data of Blazev & Lamb 1999.)

$Mg^{2+}$, low ATP and raised AMP and IMP levels, as may occur in severely fatigued muscle, may combine to decrease calcium release (Fig. 10.12).

### Inorganic phosphate

Most current interest centres on the role that $P_i$ may play in regulating calcium release. One of the most important factors regulating calcium release from the sarcoplasmic reticulum (SR) is the diffusion gradient from inside to out, which is determined by the internal stores of free calcium. It is suggested that in situations where there is a high concentration of $P_i$ (e.g. towards the end of the contractions shown in Fig. 10.7) phosphate enters the sarcoplasmic reticulum and precipitates, forming calcium phosphate and so reducing the amount of free calcium available for release (Fig. 10.13) (Fryer et al 1995, Posterino & Fryer 1998).

A problem with this explanation, and the reason why this idea was not initially widely accepted, is the fact that there is poor correlation between the time course of the increase in $P_i$ concentration and decline of force (e.g. Fig. 10.7). The major changes of $P_i$ occur in the early stages of a contraction, mirroring the decline in PCr, at a

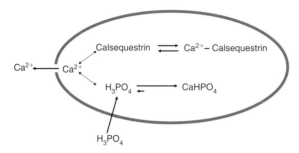

**Figure 10.13** Effects of high inorganic phosphate levels on calcium release. In fresh muscle, calcium is bound mainly to calsequestrin, from which it can rapidly dissociate and be released. In fatigued conditions, phosphate enters the sarcoplasmic reticulum and precipitates as calcium phosphate, so reducing the amount of free calcium available for release. (After Fryer et al 1995.)

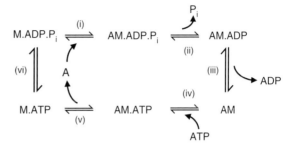

**Figure 10.14** Steps in the hydrolysis of ATP by actin and myosin during the cross-bridge cycle (see Ch. 2 for more details).

time when there is relatively little force loss. However, when force begins to decline significantly in the latter part of the contraction, there is little further change in $P_i$ levels. It is now suggested that the permeability of the SR to $P_i$ may be regulated by ATP levels, so it is not until the later stages of the contraction that the SR membrane becomes permeable to the high concentrations of $P_i$ that surround it. Alternatively, or in addition, $P_i$ may pass slowly through the SR membrane and a large proportion of the calcium may have to be sequestered before there is any noticeable effect on calcium release (for a review see Allen & Westerblad 2001).

## Influence of metabolites on cross-bridge function

The interaction of the myosin cross-bridge with actin is a series of reversible reactions, and the accumulation of products ($P_i$ and ADP) or loss of substrate (ATP) could, in theory, slow or even reverse the cycle (Fig. 10.14).

### Inorganic phosphate

Inorganic phosphate accumulates in fatigued muscle and might be expected to affect the cross-bridge cycle by slowing the transition from low- to high-force binding states—(ii) in Figure 10.14. Thus $P_i$ accumulation will decrease the relative

proportion of cross-bridges in the high-force state, resulting in a reduced isometric force. Because the next reaction (iii), the release of ADP, is rate limiting, $P_i$ accumulation would not be expected to affect shortening velocity; this is shown in Figure 10.15 (Cooke & Pate 1990). Having more cross-bridges in the low-force state would tend to increase the force obtained when stretching a muscle relative to the force of an isometric contraction; this is consistent with the data shown in Figure 9.3.

In the fatiguing protocol used by Westerblad & Allen (1993b) (Fig. 10.3) the initial, relatively small, loss of force (time point b) was not due to reduced calcium release, and probably represents the effects of raised $P_i$, decreasing the number of cross-bridges in the high-force state.

In skinned fibres, most of the effect of $P_i$ is seen with a concentration increase of between 1 and $10\,mmol\,L^{-1}$, producing a decrease in isometric force of about 35% (Fig. 10.16). Measurements of metabolites in human muscle suggest that resting levels of $P_i$ are around $8\,mmol\,L^{-1}$ (Fig. 10.7). If human muscle has the same sensitivity as the muscles studied by Cook & Pate (1990), this would imply that human muscle is partially inhibited even at rest and that increases in $P_i$ concentration with fatigue have only a minor role in affecting cross-bridge function. There may, however, be differences between fibre types: fast muscles have higher PCr and lower $P_i$ levels at rest and so fast fibres start further to the left of the relationship in Figure 10.6 and may thus be more sensitive to increases of $P_i$ in the lower range of concentration.

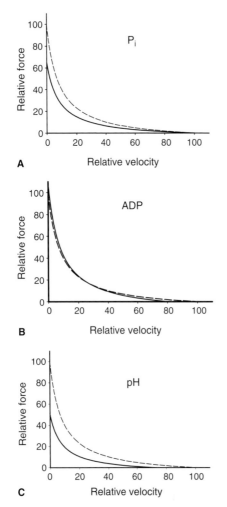

**Figure 10.15** (**A–C**) Effects of inorganic phosphate ($P_i$), ADP and pH on force–velocity relationships in skinned skeletal muscle fibres. Dashed lines are measurements in control conditions; continuous lines are with enhanced $P_i$, ADP and $H^+$, respectively. (Redrawn from the data of Cooke & Pate 1990.)

While raised $P_i$ levels may have little direct effect on cross-bridge function, they may have a major effect on calcium release and so decrease force in that way, as discussed above.

## Phosphocreatine

As with $P_i$, the time course of PCr or creatine change during activity does not match that of loss of force, and there are no suggestions that PCr or creatine have any direct effects on cross-bridge

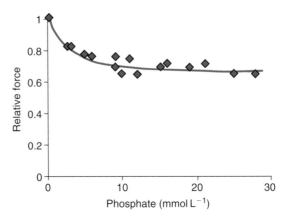

**Figure 10.16** Effects of inorganic phosphate ($P_i$) on cross-bridge force production. Skinned rabbit muscle fibres activated by calcium in the presence of different concentrations of $P_i$. (Redrawn from the data of Cooke & Pate 1990.)

function. The PCr reserves are very important to muscle and their loss indicates a considerable degree of metabolic stress, but changes in PCr concentration should be seen as an indicator of fatigue rather than its direct cause.

## ADP

An increase in ADP levels may slow reaction (iii) in Figure 10.14, increasing the back reaction and thereby increasing the mean number of cross-bridges in the active state and thus the isometric force. Slowing the release of ADP decreases the rate of cycling and will therefore reduce the maximum velocity of shortening. Such effects can be seen in Figure 10.15 for skinned fibre preparations, but the difference is small and is seen only when using unphysiological high concentrations of ADP. In living preparations ADP is maintained at low levels by exchange with PCr and substrate-level phosphorylation during glycolysis; when this fails ADP is rapidly removed by conversion to AMP and IMP. It seems unlikely that ADP has any significant effects on the force–velocity relationships of intact muscles.

## ATP

ATP is essential for cross-bridge dissociation (reaction (v), Fig. 10.14), and in its absence actin

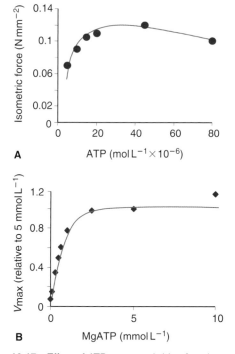

**A**

**B**

**Figure 10.17** Effect of ATP on cross-bridge function. Skinned fibre preparations activated with calcium in the presence of different concentrations of MgATP. **A**, effect on isometric force (from Cooke & Bialek 1979). **B**, effect on maximum velocity of shortening, expressed as a fraction of the velocity obtained with $5\,mmol\,L^{-1}$ ATP (Redrawn from the data of Ferenczi et al 1984.)

and myosin remain attached in the rigor complex. Consequently it might be expected that muscle would be very sensitive to fluctuations in ATP levels and that a decrease would certainly affect the maximum velocity of shortening, which is determined by the rate of cross-bridge detachment. In practice, however, cross-bridges seem to be very robust in this respect; actomyosin has a very high affinity for ATP. In skinned fibre preparations the $K_m$ for maximum velocity of shortening is $180\,\mu mol\,L^{-1}$ and isometric force is maximal at about $50\,\mu mol\,L^{-1}$ (Fig. 10.17). Given that the lowest levels of ATP that have been measured in fatigue are at least an order of magnitude greater than this, it is unlikely that low ATP concentration affects isometric force and the effect on shortening velocity is, at best, going to be small. It must be remembered, however, that even single-fibre measurements give metabolite values that are the mean for the whole cell. It is quite probable that there are compartments of the cell where ATP utilization rates are very high and may, therefore, experience lower concentrations of ATP than the average for the whole fibre.

## pH

At the point of fatigue, muscle pH can be as low as 6.5, which is towards the lower limits of what can be tolerated by most biological systems, and we may expect that many enzyme activities will be inhibited, including the interaction of actin and myosin. There is a wealth of experimental evidence showing the deleterious effects of acidosis on isolated preparations, affecting both the isometric force generated by cross-bridges and the maximum velocity of unloaded shortening (e.g. Fig. 10.15).

Similar results have been obtained with single-fibre preparations with intact membranes where it is possible to alter the internal pH by changing the concentration of carbon dioxide in the bathing medium (Curtin & Edman 1994). The large effects of pH that have been described might be thought to account entirely for the changes in the force–velocity relationship with fatigue. However, there are a number of reasons to believe that the explanation may not be this simple for mammalian muscle working in situ. While a fall in pH may play some part in the development of fatigue, it cannot be the only cause. Fatigue develops even more rapidly when glycolysis is absent and no rise in $H^+$ concentration occurs, as in isolated muscle poisoned by iodoacetate or in patients with McArdle's disease (see Fig. 10.10). In these muscles, rather than falling, pH rises slightly during the contraction as the result of PCr hydrolysis and liberation of $P_i$, which is a significant source of buffering capacity in the active muscle (Cady et al 1989a). Similar conclusions can be drawn from single-fibre experiments, where it is possible to demonstrate the consequences of fatigue in the absence of pH change (Westerblad & Allen 1992a).

More recently it has been shown that the effects of pH on muscle contractile function are very dependent on temperature. At low temperatures

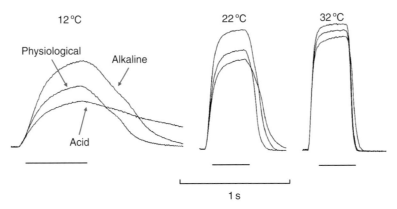

**Figure 10.18** Effects of pH on isometric force production at different temperatures. Single mouse fibres incubated at different percentages of carbon dioxide and temperatures. Note the declining influence of acidosis with increasing temperature. (From Westerblad et al 1997, with permission.)

the effect of acidity is marked, decreasing force and slowing relaxation, but at physiological temperatures pH has little effect on mammalian fibres (Fig. 10.18) (Westerblad et al 1997). Pate et al (1995) have also shown that, while low pH reduces both isometric force and maximum velocity of shortening with skinned rabbit fibres at 10°C (see Fig. 10.15C), at 30°C there is smaller effect on isometric force and shortening velocity is unaffected.

Lactate accumulation and acidosis are so widely believed to be the cause of muscle fatigue that it is appropriate to re-emphasize the point that the experimental evidence, from both human work and isolated mammalian preparations at physiological temperatures, clearly shows that $H^+$ accumulation is not the only, or even the major, cause of fatigue during high-intensity exercise.

Having emphasized the point that pH is not as important as widely believed, it has to be noted that there is still some effect of acidity to be seen at 32°C (Fig. 10.18). An effect of pH can also be seen in the recovery phase following fatiguing exercise in human muscle where pH recovers more slowly than the phosphorus metabolites and, for the same PCr (and by implication $P_i$, ATP and ADP) concentration, the more acid muscle generates less force and remains slower (Fig. 10.19). This observation implies that, in addition to the pH-independent mechanisms discussed

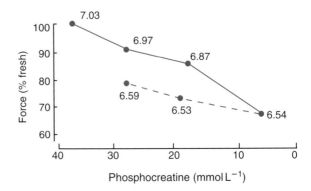

**Figure 10.19** Force and pH during fatigue and recovery. Human first dorsal interosseous muscle fatigued (see Fig. 10.7) and then followed during recovery. The solid line shows the extent of metabolic change, and the dotted line the recovery phase. Values denote the intracellular pH. Note that for similar values of phosphocreatine less force was generated during recovery when the pH was low. (Data from Cady et al 1989a.)

above, there is also a pH-dependent mechanism that plays some part in fatigue.

It is not clear which stage of the cross-bridge cycle may be affected by $H^+$ in fatigue. The action of $H^+$ on shortening velocity (at lower temperatures) suggests it may slow ADP dissociation, the binding of ATP or the subsequent dissociation of actin and myosin. The action of $H^+$ on decreasing isometric force could be by increasing the reverse reaction—(ii) in Figure 10.14; an alternative might be to slow the hydrolysis of ATP bound to myosin—(v) in Figure 10.14. This would have the effect of slowing the re-priming of myosin heads and thus reducing the number of cross-bridges available for attachment. The issue might be resolved by stretching the muscle. If there were fewer cross-bridges available then the force obtained during stretch would be in proportion to the isometric force, whereas if there was an increase in the proportion of cross-bridges in a low-force state then a relatively large increase in force during stretch would be obtained. This latter prediction is very much what is seen (see Fig. 9.13).

## SUMMARY

The reader may be forgiven for being confused at this point. The importance of changes in force–velocity characteristics has been emphasized, and these indicate changes in cross-bridge function. These changes are, in turn, almost certainly the result of some change in metabolite concentration. However, there is no clear indication as to which metabolite is important. Phosphate accumulation probably affects only force, and not shortening velocity; ADP is tightly regulated and probably does not accumulate; and ATP levels do not appear to fall far enough to affect cross-bridge function. $H^+$ accumulation can have serious effects on cross-bridge function, but there is evidence from several sources that it plays a minor role in fatigue of mammalian muscle at physiological temperatures. Most workers in the field are resorting to arguments about compartmentalization to explain the apparent contradictions. A reduction of ATP or an increase in ADP levels in a small compartment could have effects on cross-bridge or SR function without being apparent in measurements of the whole fibre metabolite content. Alternatively, or in addition, fatigue effects may be due to a combination of changes. Most work to date has examined the effect of changing one substance at a time. A slightly reduced ATP and a small increase in ADP concentration in the presence of a reduced pH may be much more effective in influencing cross-bridge function than any one change taken alone.

REFERENCES AND FURTHER READING

Allen D G, Westerblad H 2001 Role of phosphate and calcium stores in muscle fatigue. Journal of Physiology 536: 657–665

Allen D G, Westerblad H, Lännergren J 1997 The role of ATP in the regulation of intracellular $Ca^{2+}$ release in single fibres of mouse skeletal muscle. Journal of Physiology 498: 587–600

Allen D G, Lännergren J, Westerblad H 1999 The use of caged adenine nucleotides and caged phosphate in intact skeletal muscle fibres of the mouse. Acta Physiologica Scandinavica 166: 341–347

Bigland-Ritchie B, Kukulka C G, Lippold O C J, Woods J J 1982 The absence of neuromuscular junction transmission failure in sustained maximal voluntary contractions. Journal of Physiology 330: 265–278

Blazev R, Lamb G D 1999 Low [ATP] and elevated $[Mg^{2+}]$ reduce depolarization-induced $Ca^{2+}$ release in rat skinned muscle fibres. Journal of Physiology 520: 203–215

Cady E B, Jones D A, Lynn J, Newham D J 1989a Changes in force and intracellular metabolites during fatigue of human skeletal muscle. Journal of Physiology 418: 311–325

Cady E B, Elshove H, Jones D A, Moll A 1989b The metabolic causes of slow relaxation in fatigued human skeletal muscle. Journal of Physiology 418: 327–337

Cooke R, Bialek W 1979 Contraction of glycerinated muscle fibres as a function of MgATP concentration. Biophysical Journal 48: 241–258

Cooke R, Pate E 1990 The inhibition of muscle contraction by the products of ATP hydrolysis. In: Taylor A W et al (eds) Biochemistry of exercise VII, pp. 59–72. Champaign, Illinois: Human Kinetics Books

Curtin N A, Edman K A P 1994 Force–velocity relation for frog muscle fibres: effects of moderate fatigue and of intracellular acidification. Journal of Physiology 475: 483–494

Donaldson S K B, Hermansen L 1978 Differential, direct effects of $H^+$ on $Ca^{2+}$-activated force of skinned fibres from soleus, cardiac and adductor magnus muscles of rabbits. Pflügers Archiv 376: 55–65

De Haan A, de Ruiter C J, van der Woude L H V, Jongen P J H 2000 Contractile properties and fatigue of quadriceps muscles in multiple sclerosis. Muscle and Nerve 23: 1534–1541

Eberstein A, Sandow A 1963 Fatigue mechanisms in muscle fibres. In: Gutman E, Hnik P (eds) The effect of use and disuse on neuromuscular functions, pp 515–526. Prague: Czechoslovakian Academy of Science

Edwards R H T, Hill D K, Jones D A 1975 Metabolic changes associated with slowing of relaxation in fatigued mouse muscle. Journal of Physiology 251: 287–301

Ferenczi M A, Goldman Y E, Simmons R M 1984 The dependence of force and shortening velocity on substrate concentration in skinned muscle fibres from Rana temporaria. Journal of Physiology 350: 519–543

Fryer M W, Owen V J, Lamb G B, Stephenson D G 1995 Effects of creatine phosphate and $P_i$ on $Ca^{2+}$ movements and tension development in rat skinned skeletal muscle fibres. Journal of Physiology 482: 123–140

Hnik P, Holas M, Krekule I et al 1976 Work-induced potassium changes in skeletal muscle and effluent venous blood assessed by liquid ion-exchanger microelectrodes. Pflügers Archiv 362: 84–95

Jones D A 1996 High and low frequency fatigue revisited. Acta Physiologica Scandinavica 156: 265–270

Jones D A, Bigland-Ritchie B 1986 Electrical and contractile changes in muscle fatigue. In: Saltin B (ed.) Biochemistry of exercise VI, pp 337–392. International Series on Sports Sciences, vol 16. Champaign, Illinois: Human Kinetics

Jones D A, Sacco P 1988 Failure of activation as the cause of fatigue in isolated mouse skeletal muscle. Journal of Physiology 419: 75P

Karatzaferi C, de Haan A, Ferguson R A, van Mechelen W, Sargeant A J 2001a Phosphocreatine and ATP content in human single muscle fibres before and after maximum dynamic exercise. Pflügers Archiv. European Journal of Physiology 442: 467–474

Karatzaferi C, de Haan A, van Mechelen, W, Sargeant A J 2001b Metabolic changes in single human muscle fibres during brief maximal exercise. Experimental Physiology 86: 411–415

Kernell D, Lind A, van Diemen A B J P, de Haan A 1995 Relative degree of stimulation-evoked glycogen degradation in muscle fibres of different type in rat medial gastrocnemius. Journal of Physiology 484: 139–153

Lamb G D, Stephenson D G 1991 Effects of $Mg^{2+}$ on the control of $Ca^{2+}$ release in skeletal muscle fibres of the toad. Journal of Physiology 434: 507–528

Lamb G D, Stephenson D G 1994 Effects of intracellular pH and $[Mg^{2+}]$ on excitation–contraction coupling in skeletal muscle fibres of the rat. Journal of Physiology 478: 331–339

Lännergren J, Westerblad H 1986 Force and membrane potential during and after fatiguing, continuous high frequency stimulation of single *Xenopus* muscle fibres. Acta Physiologica Scandinavica 128: 359–368

Lännergren J, Westerblad H 1987 Action potential fatigue in single skeletal muscle fibres of *Xenopus*. Acta Physiologica Scandinavica 129: 311–318

Lee J A, Westerblad H, Allen D G 1991 Changes in tetanic and resting $[Ca^{2+}]_l$ during fatigue and recovery of single muscle fibres of *Xenopus laevis*. Journal of Physiology 433: 307–327

Nielsen O B, Clausen T 2000 The $Na^+/K(^+)$-pump protects muscle excitability and contractility during exercise. Exercise and Sport Sciences Reviews 28: 159–164

Pate E, Bhimani M, Franks-Skiba K, Cooke R 1995 Reduced effect of pH on skinned rabbit psoas muscle mechanics at high temperatures: implications for fatigue. Journal of Physiology 486: 689–694

Posterino G S, Fryer M W 1998 Mechanisms underlying phosphate-induced failure of $Ca^{2+}$ release in single skinned skeletal muscle fibres of the rat. Journal of Physiology 512: 97–108

Ruff R L, Simiocini L, Stühmer W 1987 Comparison between slow sodium channel inactivation in rat slow- and fast-twitch muscle. Journal of Physiology 383: 339–348

Sjögaard G, McComas A J 1995 Role of interstitial potassium. Advances in Experimental Medicine and Biology 384: 69–80

Sjögaard G, Adams R P, Saltin B 1985 Water and ion shifts in skeletal muscle of humans with intense dynamic knee extension. American Journal of Physiology 248: R190–R196

Westerblad H, Allen D G 1991 Changes in myoplasmic calcium concentration during fatigue in single mouse muscle fibres. Journal of General Physiology 98: 615–635

Westerblad H, Allen D G 1992a Changes in intracellular pH due to repetitive stimulation of single fibres from mouse skeletal muscle. Journal of Physiology 449: 49–71

Westerblad H, Allen D G 1992b Myoplasmic free $Mg^{2+}$ concentration during repetitive stimulation of single fibres from mouse skeletal muscle. Journal of Physiology 453: 413–434

Westerblad H, Allen D G 1993a The influence of intracellular pH on contraction, relaxation and $[Ca^{2+}]_i$ in intact single fibres from mouse muscle. Journal of Physiology 466: 611–628

Westerblad H, Allen D G 1993b The contribution of $[Ca^{2+}]_i$ to the slowing of relaxation in fatigued single fibres from mouse skeletal muscle. Journal of Physiology 468: 729–740

Westerblad H, Bruton J D, Lännergren J 1997 The effect of intracellular pH on contractile function of intact, single fibres of mouse muscle declines with increasing temperature. Journal of Physiology 500: 193–204

Westerblad H, Allen D G, Bruton J D, Andrade F H, Lännergren J 1998 Mechanisms underlying the reduction of isometric force in skeletal muscle fatigue. Acta Physiologica Scandinavica 162: 253–260

Westerblad H, Allen D G, Lännergren J 2002 Muscle fatigue: lactic acid or inorganic phosphate the major cause? News in Physiological Sciences 17: 17–21

Yoshida T, Watari H 1992 Noninvasive and continuous determination of energy metabolism during muscular contraction and recovery. In: Sato Y et al (eds) Medicine and sport science, vol 37: Integration of medical and sport sciences, pp 364–373. Basel: Karger

## QUESTIONS TO THINK ABOUT

1. If lactate accumulation and intracellular acidosis are not the main cause of fatigue, why does lactate concentration usually correlate so well with the onset of changes in performance?

2. What would be the influence on cross-bridge cycling (and hence muscle function) if the concentrations of the following metabolites were separately increased: ATP, ADP, $P_i$ (inorganic phosphate) and $H^+$?

3. The cross-bridge is one site at which metabolites can affect function. What other sites within the muscle fibre are also sensitive to changes in metabolite concentrations? Are all fibre types affected in similar ways by such changes?

4. Metabolite concentrations may not be uniform throughout the muscle fibre. Indicate where in the cell and under what conditions differences can occur in the ATP/ADP ratio.

5. What is meant by the term 'excitation–contraction coupling'? What structures are involved and how may they be affected in fatigue?

6. What experimental evidence is there that calcium release may fail in fatigued muscle?

7. What factors contribute to the loss of power seen in acutely fatigued muscle? If $V_{max}$ changes, what does this tell us about cross-bridge function?

8. Discuss the causes of fatigue in patients with genetic defects in enzymes of the glycolytic pathway (e.g. McArdle's disease).

9. Why are the T tubules a potential site of electrical failure?

10. What protective mechanisms are there that prevent or minimize electrical failure during high-intensity exercise?

# 11

# Fatigue during endurance exercise

Chapters 9 and 10 discussed the possible causes of fatigue during high-intensity exercise, typically lasting for 40–60 s and corresponding roughly to sprinting 400 m. The characteristic of this type of exercise is that the energy demands are greater than the supply and there is a run-down of the high-energy reserves of the cell and an accumulation of lactate and $H^+$.

For events over 800–1500 m, a substantial part of the race must be run with the body in a steady state, that is, oxidative metabolism must keep up with the demands of exercise and ventilation, and heart rate and oxygen uptake should all be at relatively constant (albeit high) levels.

The first topics for discussion are the metabolic factors that determine the level of steady-state exercise, and the question of how long this level can be sustained. Steady-state exercise in routine fitness testing usually means keeping heart rate and oxygen uptake constant for the 3–4-min stages, but for events such as marathons, triathlons and cycle road races a high and fairly constant power output is required for several hours.

## FACTORS THAT DETERMINE MAXIMUM OXYGEN UPTAKE

The transport of oxygen from the inspired air to the muscles and its role in oxidative metabolism involves a series of events, which may be divided into *delivery* and *utilization*. On the delivery side there are factors such as lung volume and perfusion, cardiac output and blood pressure, blood volume, haematocrit and haemoglobin content. At the level of the muscle, and still arguably classified as 'delivery', there is the capillary density, the dissociation curve of haemoglobin and the myoglobin content of the fibres, which will determine the rate of oxygen diffusion into the fibre. One further factor that assists in the delivery of oxygen (and the diffusion outwards of carbon dioxide) is the diameter of the muscle fibre: the smaller the fibre, the shorter the diffusion distance. In terms of utilization, the primary limiting factor is the muscle fibre content of mitochondria and then the quantity and quality of the fuel that is oxidized.

The balance between central factors (delivery) and peripheral factors (utilization) in setting the maximum aerobic capacity ($VO_{2max}$) has long been debated and can be shown to be one or the other depending on the circumstances. In subjects with large muscular legs and a relatively small cardiac output (such as power athletes), the maximum oxygen uptake that can be achieved when exercising both legs is less than twice the maximum uptake achieved with just one leg, suggesting that the central pump cannot supply sufficient oxygen to match the capacity of both legs. In contrast, endurance athletes, with large hearts and lungs and relatively small leg muscles, can supply enough oxygen to meet the needs of both legs working maximally; the $VO_{2max}$ of two-leg exercise being twice that of just one leg. As with all arguments about weak links in chains, it is likely that in 'normal' circumstances the various components all have a similar capacity, on the one hand to deliver and on the other to utilize oxygen, and it is only in extreme cases of highly trained endurance or power athletes that one aspect outweighs the other.

The factors determining the limits of oxygen delivery and carbon dioxide removal are dealt with in all textbooks of respiratory physiology, so we concentrate here on the metabolic limitations within the muscle. However, we also consider the part that neural control, or central fatigue, may play in prolonged exercise, especially when undertaken in hot and humid conditions.

## Muscle fibre size

Successful endurance athletes have a characteristic ectomorphic body shape and composition, in that they tend to be thin with long limbs, narrow shoulders and rather small muscle development. It is likely that the small muscle size indicates relatively small muscle fibres, and the shorter distances between capillary and the interior of the muscle fibre facilitates the diffusion of oxygen and carbon dioxide. The heart is the supreme example of an endurance muscle. Although cardiac muscle has a high mitochondrial content, it is not very much greater than that of slow type 1 skeletal muscle fibres and yet the heart is capable of maintaining a high power output, not just for 1–2 h but for 70–80 years, something that even the most highly trained and adapted skeletal muscle cannot do. The diameter of a cardiac muscle fibre is about one-fifth that of a skeletal muscle fibre and it is probably this factor that accounts for the differences in endurance. Endurance training of skeletal muscle tends to cause a reduction in fibre cross-sectional area, but the skeletal fibres never become as small as cardiac fibres. Type 1 oxidative fibres are significantly smaller than the type 2 fibres in many animals, although in human muscle the fibres are of similar cross-sectional area.

One other attribute of the ectomorphic build of endurance athletes is that they lack the heavy upper body musculature of power athletes, and consequently have minimal amounts of superfluous weight to carry when running or cycling.

## Mitochondria

There are large differences in the content of oxidative enzymes between different fibre types (see Figs 4.3 & 4.4) and this can also be seen in EM sections of muscle (Fig. 11.1). Elite endurance athletes have a predominance of the slow type 1 fibres and consequently their oxidative capacity is higher than that of the normal population. In addition, endurance training can approximately double the content of mitochondria and the constituent enzymes of the tricarboxylic acid cycle and electron transport chain. The major change with such training is seen in the type 2 fibres.

The main increase in mitochondrial density with training occurs in the subsarcolemmal population. Large clumps of mitochondria are seen around the periphery of highly oxidative fibres in

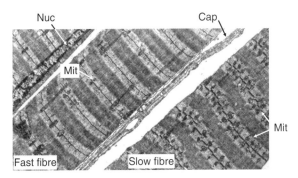

**Figure 11.1** Low-power electron micrograph of two adjacent fibres, one fast, the other slow. The fibre types can be distinguished by the width of their Z lines (wider in the slow fibres) and content of mitochondria. Nuc, muscle fibre nucleus; Cap, blood capillary; Mit, mitochondria.

endurance animals, and it is interesting that similar accumulations of mitochondria are seen in human muscle diseases with defective mitochondria (ragged red fibres; see Fig. 15.7). In the disease state, the mitochondria proliferate in an apparent attempt to compensate for their defective function. Presumably there is a similar stimulus in normal subjects during training when the level of oxidative metabolism is not sufficient to cope with the demand. There are obvious advantages to having mitochondria close to the surface membrane, as this will minimize the diffusion distances for oxygen.

Oxidative fibres have a high myoglobin content and endurance training can increase the content by up to 80%, the major part of the increase being in the faster 2a and 2x fibres. Diving mammals, such as whales and seals, have exceptionally high levels of muscle myoglobin, making their meat appear very dark. In these animals the myoglobin acts as a local store of oxygen, facilitating their long dives. In land animals the myoglobin may constitute a small local store, but probably its main function is to reduce oxygen tension within the muscle fibre, thus promoting dissociation of oxygen from haemoglobin in red blood cells in adjacent capillaries (Fig. 11.1) and its diffusion into the fibre.

## Fuel

For a given mitochondrial content and supply of oxygen, the maximum rate of ATP production depends on the fuel being used, as discussed in Chapter 6.

The maximum power sustained during sprinting is determined by the rate at which ATP can be regenerated from phosphocreatine and glycolysis and, in the fastest individuals, allows a running speed of approximately $10\,m\,s^{-1}$. The fastest rates of oxidative metabolism occur when using carbohydrate as a fuel and, in general, this allows running speeds of around $5\,m\,s^{-1}$, the speed at which events ranging from 3000 m to the marathon are run. Oxidation of fat is slower than that of carbohydrate, and by itself can sustain exercise at a running speed of about $2\,m\,s^{-1}$. Two metres per second is a fast shuffle, the pace at which 24- and 100-h races are run, or shuffled. The reason why fat oxidation is so slow is not fully understood. A number of limiting steps have been suggested, including the hydrolysis of triglycerides, the capacity of albumin to carry fatty acids to the muscle, their uptake across the surface membrane and subsequent transport across the mitochondrial membranes, before entering the spiral of β-oxidation (Jeukendrup et al 1998a,b).

## FOR HOW LONG CAN A HIGH $V_{O_2}$ BE MAINTAINED?

The conventional answer to this question is that a high $V_{O_2}$ (around 80%) can be maintained for as long as there is carbohydrate available for oxidation, as fatty acid oxidation alone can sustain a work rate of only about 50% $V_{O_2max}$. There are two ways of prolonging this phase of carbohydrate utilization: by increasing the muscle reserves of glycogen or by sparing its use by maximizing the oxidation of fatty acids.

### Muscle glycogen reserves

Figure 11.2 shows the rate at which muscle glycogen is used during exercise at different intensities. If an athlete were to set off at a competitive marathon pace ($5\,m\,s^{-1}$, 80% $V_{O_2max}$) with the starting level of muscle glycogen shown in Figure 11.2, the carbohydrate stores would be exhausted after about 75 min. At this point there

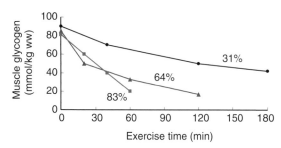

**Figure 11.2** Muscle glycogen depletion during exercise at three different intensities (%$VO_{2max}$). (Redrawn from the data of Gollnick et al 1974.)

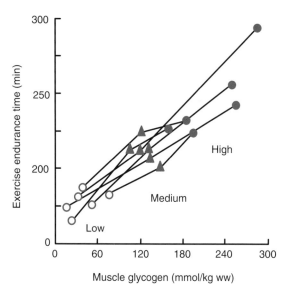

**Figure 11.3** Endurance as a function of initial muscle glycogen. Subjects manipulated their muscle glycogen by taking a diet with a low, medium or high carbohydrate content. Values for muscle glycogen are given as millimolar glucosyl units per kilogram wet muscle. (Redrawn from the data of Bergström et al 1967.)

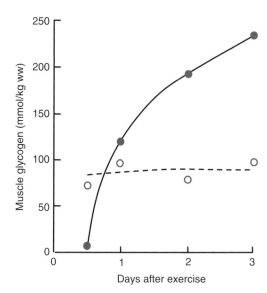

**Figure 11.4** Resynthesis of muscle glycogen after exercise. Subjects depleted one leg of glycogen and biopsies were taken from both legs on subsequent days. Solid line and symbols, depleted leg; dashed line and open symbols, control leg. Values for muscle glycogen are given as millimolar glycosyl units per kilogram wet muscle. (Redrawn from the data of Bergström & Hultman 1966.)

would be no alternative but to oxidize fat, which can support power output at only about half the value obtained from carbohydrate. Consequently running speed would fall to a fast shuffle, which is all that can be maintained at 50% $VO_{2max}$. This sudden change of pace is known as 'hitting the wall' and is characteristic of ill-prepared athletes running marathons.

Muscle glycogen stores are responsive to dietary carbohydrate, and this can have a considerable influence on endurance performance. Figure 11.3

shows subjects whose muscle glycogen was manipulated by varying the carbohydrate content of their diet. Endurance times varied between 1 and 3 h in most subjects, and up to 5 h in one subject, in proportion to their muscle glycogen reserves at the start of exercise.

Bergström & Hultman (1966) were the first to notice the phenomenon of 'super-compensation' during an investigation of the rate of glycogen resynthesis following exercise. Their subjects exercised one leg to exhaust the muscle glycogen while the other leg acted as a control. It was found that, having depleted the test leg of glycogen, muscle levels were restored in less than a day. However, the surprising finding was that on the second and third days the glycogen level was 150–200% that of the control muscles (Fig. 11.4); this would have obvious benefits in terms of endurance performance. Glycogen synthase, the enzyme regulating glycogen synthesis, is activated by exercise, and it is particularly important to take oral carbohydrate as soon after exercise as possible to make use of this fact and maximize glycogen resynthesis.

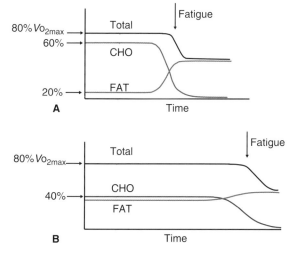

**Figure 11.5** Fatty acid oxidation sparing muscle glycogen and prolonging endurance. **A**, carbohydrate (CHO) makes the major contribution to total energy production and is used up rapidly. **B**, fatty acid oxidation accounts for half the total required energy production, and fatigue is postponed.

## Conservation of glycogen stores

Exercise at 80% $VO_{2max}$ can be achieved by oxidizing carbohydrate, but if this is the only source of energy muscle glycogen will be used within 60–90 min. However, if fatty acid oxidation is used to provide some of the energy, perhaps equivalent to 40% $VO_{2max}$, then glycogen will have to provide only the remaining 40% and will thus last for twice as long as when it was the sole source of energy (Fig. 11.5). Increasing the utilization of fatty acids is therefore a way of conserving muscle glycogen stores, which are essential for high workloads. Endurance athletes have the ability to oxidize fats at a faster rate than untrained subjects and training has been shown to approximately double the quantity of fatty acid used during 1 h of exercise at 60% $VO_{2max}$.

Animal studies have shown that increasing the availability of circulating free fatty acids by giving a lipid meal and injection of heparin (which activates lipoprotein lipase, promoting uptake of fatty acids) spares muscle glycogen during exercise (Costill et al 1977, Rennie et al 1976). It has been widely assumed that training human subjects causes a greater or earlier mobilization of fatty acids, making these available to the muscle.

However, there is little difference between levels of circulating free fatty acids during exercise in trained and untrained subjects (Kiens et al 1993) and it would appear that the major difference is in the uptake of fatty acids into the muscle. In addition, it is apparent that the intramuscular stores of triglyceride are also an important source of fuel for the trained athlete.

One response that might stimulate fatty acid metabolism in the trained athlete is for there to be a more rapid or greater hormonal response to exercise, for example more adrenaline or growth hormone, to activate *hormone-stimulated lipase*, which splits fatty acids from glycerol. In fact, quite the reverse occurs: trained endurance athletes have a smaller hormonal response and it is the inexperienced athletes who are nervous and have high circulating levels of adrenaline at the start of exercise (Bloom et al 1976). If, therefore, there is no increased hormone response, there might be an increased receptor density or affinity, much as occurs with insulin receptors in response to training. However, here again, the evidence is conflicting and currently there is no clear mechanism whereby training increases fatty acid metabolism. Quite possibly there is a series of relatively small changes: increased mobilization, increased transport of fatty acids into the fibre, increased numbers of mitochondria.

When energy production can no longer match expenditure, either because muscle glycogen is exhausted or the energy demands of the race have increased with a hill, a break away from the pack or a sprint for the finish, the reserves of phosphocreatine will be called on and glycolysis will be stimulated. All the mechanisms of fatigue, discussed in Chapters 9 & 10, associated with high-intensity exercise, will come into play. Calcium release will decrease and cross-bridge function will slow. In this respect endurance exercise can be seen as a long prelude to a metabolic crisis. The duration of this prelude depends on the balance between rates of delivery and utilization of energy and, as described above, training and diet can make major differences to endurance performance.

The effects of diet are shown in Figure 11.3, but the results are unusual in one respect. The subjects were reported to have exercised at 80%

$V_{O_{2max}}$ for about 3 h and, in one case, for up to 5 h after the high carbohydrate diet. This is a considerable achievement, because most fit young subjects who are accustomed to exercise can manage about 60 min at this intensity, despite being well fed and showing every sign that they are exercising in a steady state. Most people who are not accustomed to this type of activity will complain loudly after about 20 min.

## CENTRAL FATIGUE DURING PROLONGED EXERCISE

It is a common observation that during prolonged exercise the point of fatigue frequently does not coincide with a metabolic crisis developing in the muscle. Even if subjects are prepared to push themselves until their muscles are depleted of glycogen, the sensations of fatigue, or exertion, increase steadily throughout the exercise and bear little relationship to heart rate, ventilation or circulating lactate levels. This has led to speculation about a central, as opposed to a peripheral, mechanism of fatigue during prolonged exercise.

At present, central fatigue is inferred by default, that is, by demonstrating that peripheral muscle function is preserved at a time when overall voluntary activity is failing. It is difficult to make meaningful measurements of muscle function while running or cycling, but the few attempts that have been made, either of isometric strength immediately at the end of exercise or of maximum sprint power, have shown little or no decrease in peripheral muscle performance. Indirect evidence comes from observations that blood lactate concentration is not particularly high at the point of fatigue, the subjects are not hypoglycaemic, muscle glycogen is not exhausted, and heart rate and ventilation are not maximal (e.g. Pitsiladis & Maughan 1999).

### Tryptophan and central fatigue

In the past 15 years considerable attention has been focused on serotonergic pathways in the brain and the influence that circulating levels of tryptophan have on their function.

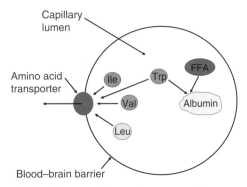

**Figure 11.6** Transport of tryptophan across the blood–brain barrier. Tryptophan competes with the branched-chain amino acids isoleucine, valine and leucine for the same transporter. In addition, the majority of tryptophan is bound to albumin, so reducing the quantity of free tryptophan available for transport. Free fatty acids (FFA) compete for the same binding site on albumin.

Serotonin is a neurotransmitter involved in regulating mood and motivation, and drugs commonly used in the treatment of depression modify the function of serotonergic pathways. Monoamine oxidase inhibitors (tricyclics) block the breakdown of serotonin, whereas the selective serotonin reuptake inhibitors (SSRIs), as their name implies, block the reuptake of serotonin across the presynaptic membrane. Both drugs tend to increase the quantity of serotonin remaining in the synapse and lead to a downregulation of receptor numbers and, ultimately, activity of the serotonergic pathways. This is, however, a medium- to long-term effect, and in the short term SSRIs can exacerbate feelings of depression by stimulating these same pathways. Serotonin is synthesized from the amino acid tryptophan; the rate-limiting step is the supply of tryptophan. Tryptophan is at a low concentration in plasma and its movement across the blood–brain barrier is restricted by two factors (Fig. 11.6).

Tryptophan has to compete with branched-chain amino acids (BCAAs) for the transporter and, in addition, most of the tryptophan in the blood is bound to albumin. Consequently the availability of tryptophan in the brain, as a precursor of serotonin, is limited. Newsholme (Newsholme & Blomstrand 1996) realized that this situation might change with prolonged exercise. The first thing to happen is that circulating free fatty acids increase and these displace tryptophan from the albumin binding sites. Second, with prolonged exercise, the plasma BCAAs may decrease as they are taken up by the liver for gluconeogenesis. The combination of these two effects increases the flux of tryptophan across the blood–brain barrier and, in theory, increases the synthesis of serotonin. This acute increase in serotonin concentration was suggested by Newsholme to be the origin of feelings of fatigue and the failure of performance at a time when subjects appear to be in a steady state as far as measures of cardiovascular and respiratory function are concerned.

One of the attractions of Newsholme's theory is that it suggests ways in which endurance performance may be improved and, incidentally, the theory tested. The increased transport across the blood–brain barrier suggested to occur during endurance exercise might be prevented in two ways. The concentration of BCAAs could be increased by taking supplements, and the liberation of free fatty acids might be suppressed or delayed by taking carbohydrate during the exercise (Davis 1996). Although there is some limited evidence from field studies that this type of intervention might be useful (Newsholme & Blomstrand 1996), more rigorous laboratory-based studies have failed to demonstrate any benefits (e.g. van Hall et al 1995) (Fig. 11.7). Supplements containing high tryptophan levels did not decrease performance, and neither did high or low concentrations of BCAAs improve performance.

## Central serotonergic and dopaminergic pathways

The lack of support for Newsholme's ideas concerning the role of tryptophan and central fatigue is sometimes taken as evidence either that central fatigue is not a major factor in endurance exercise or that serotonin is not involved. Neither is necessarily true. What the work with BCAAs has shown is that tryptophan uptake into the brain during exercise is probably not the mechanism whereby serotonergic pathways are activated and central fatigue is generated. Great interest

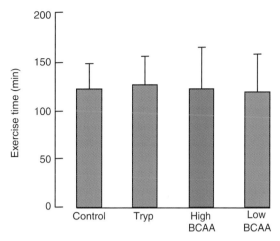

**Figure 11.7** Effect of manipulating plasma free tryptophan concentration on exercise endurance time. Subjects exercise at 70–75% maximum aerobic power output when taking either tryptophan (Tryp) and high or low supplements of branched-chain amino acids (BCAA). (Redrawn from the data of Van Hall et al 1995.)

continues in central fatigue and the role of serotonin in regulating mood and motivation.

There are other lines of evidence to suggest that serotonergic and/or dopaminergic pathways are important in regulating endurance exercise. Manipulating brain serotonin and dopamine activities in rats with appropriate agonists and antagonists has an effect on endurance capacity. In general, increasing dopaminergic activity and reducing that of serotonergic pathways increases endurance, while increasing serotonergic activity decreases exercise tolerance. There are few such drugs that can be used with human subjects, but the limited data available suggest that increasing serotonergic activity has a similar effect in reducing endurance (Marvin et al 1997, Wilson & Maughan 1992).

One further line of evidence suggests the involvement of serotonergic and/or dopaminergic systems in fatigue during endurance exercise. One of the characteristic responses to exercise is the release of stress hormones from the anterior pituitary gland. These include adrenocorticotrophic hormone, growth hormone and prolactin, the release of which is controlled by the hypothalamus. Prolactin release is stimulated by hypothalamic serotonergic activity and inhibited by dopaminergic activity, and the fact that circulating levels increase during exercise clearly indicates a change in the activity of central neural pathways (Fig. 11.8).

Demonstrating that there are central changes associated with fatiguing endurance exercise is important, but it begs two questions. The first is what aspect of endurance exercise precipitates the central changes? The second question concerns the way in which these central changes translate into a change in motivation or ability to continue exercise. One answer to the first question may come from the observation that endurance exercise is much more difficult when undertaken in hot and humid conditions.

## EXERCISE IN THE HEAT

There is a wide variation in the extent to which high ambient temperature affects different people but, typically, endurance times are reduced by about a third when exercising at 35°C compared with a more comfortable 18°C (Fig. 11.9).

Exercise in the heat, compared with cool conditions, places greater demands on skeletal muscle for two reasons. The first is that muscle temperature will be raised, perhaps by 2–3°C, thereby increasing the metabolic rate and leading to a more rapid depletion of muscle glycogen. The second reason is that the requirements of thermoregulation redirect blood flow to the skin and away from working skeletal muscles so that they are deprived of essential oxygen (Fink et al 1975).

However, in practice, neither of these changes appears to be critical. Muscles have a higher glycogen content at the point of fatigue after exercise in the heat than when exercise is undertaken in cool conditions (Pitsiladis & Maughan 1999), showing that the more rapid utilization is not the problem. In addition, leg blood flow is affected very little by exercise in the heat, and perfusion decreases significantly only when the subjects are dehydrated. Even in these circumstances, oxygen extraction increases and compensates for the reduced blood flow (Gonzalez-Alonso et al 1999a).

Subjects tend to stop exercising when their deep body temperature reaches around 40°C.

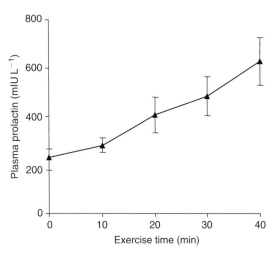

**Figure 11.8** Plasma prolactin levels during prolonged exercise. Subjects exercised at 75% $V_{O_{2max}}$ at an ambient temperature of 35°C.

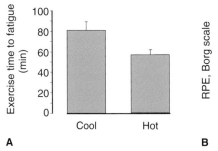
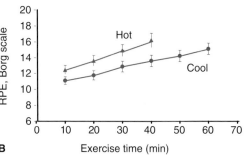

**Figure 11.9** Effect of ambient temperature on endurance performance and perception of exertion. **A**, endurance times; **B**, rating of perceived exertion (Borg scale) during exercise at 75% $V_{O_{2max}}$ in either cool (18°C) or hot (37°C) ambient conditions.

This observation was first made in relation to a study of heat acclimation. Subjects who exercised every day for 10 days in a sauna increased their exercise tolerance from about 40 min to 70 min, but stopped when their core temperature reached the value that they had achieved before acclimation. What had changed was the subjects' starting core temperature and, to a lesser extent, their ability to dissipate heat during the exercise (Fig. 11.10A). The suggestion was made that the limiting factor in this type of endurance activity was some aspect of body temperature per se (Nielsen et al 1993). This idea was substantiated by a further study in which the initial core temperature was manipulated up or down and it was shown that endurance was determined by the time it took for the core temperature to reach a critical level (Fig. 11.10B). There may well be distinct advantages to cooling the body before the start of endurance events, especially if the ambient conditions are hot and humid and core temperature may not stabilize during the event. Against this has to be set the disadvantage of reduced power output by cooler muscles (see Fig. 8.4), although for endurance events this may not be a serious problem.

The conclusion of these observations is that some aspect of body temperature affects exercise performance and there appears to be a critical temperature above which subjects are very reluctant to continue exercising.

The importance of an upper limiting temperature for exercise is evident: above about 41°C homeostatic functions begin to fail, sweating stops, blood pressure drops and multiple organ failure can ensue as the circulation collapses. The disinclination to exercise at high temperatures therefore acts as a protective reflex and where this reflex is overridden, for instance in military action and training, or with party-goers or athletes taking central stimulants (see below), severe heat illness, injury and occasionally death may ensue.

## Central mechanisms affected by raised body temperature

The fact that body temperature limits exercise through a central mechanism raises the second question concerning the nature of this mechanism. One clue comes from the observation that the secretion of prolactin during exercise is related to increasing body temperature (Fig. 11.11) and is accentuated when the ambient temperature is raised. Raised body temperature, in the absence of exercise, will also release prolactin from the pituitary (Brisson et al 1986).

The pituitary secretion of prolactin is under the control of the hypothalamus, which is also the thermoregulatory centre. There are warm and cool receptors in the preoptic area and these signals are integrated with information from peripheral temperature receptors to produce the appropriate response: vasodilatation and sweating in the heat, shivering in the cold. It seems likely, therefore, that the same hypothalamic pathways also regulate the hormonal response to heat generated by exercise.

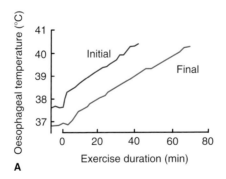

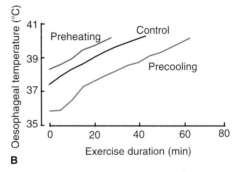

**Figure 11.10** Body temperature and exercise duration. **A**, subject exercising at 50% $\dot{V}O_{2max}$ before and 10 days after heat acclimation. (Data from Nielsen et al 1993.) **B**, subjects exercising under similar conditions but with preheating or cooling to raise or lower core temperature at the start of exercise. (Redrawn from the data of Gonzalez-Alonso et al 1999b.)

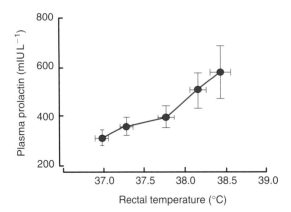

**Figure 11.11** Plasma prolactin in relation to rectal temperature during exercise. Subjects exercised at 75% $V_{O_{2max}}$ at an ambient temperature of 35°C.

The fact that perception of exertion increases with body temperature and circulating prolactin levels suggests that there is an integrated response to exercise in the heat. The first output is thermoregulation, which increases heat loss; the second is the neuroendocrine response, involving prolactin and probably growth hormone; and the third is a behavioural response decreasing the drive to exercise, which has the effect of reducing the rate of heat production. This behavioural response is manifest as an increasing perception of exertion and a desire to move to a cool environment.

The hypothalamus is probably the key integrating centre in the brain, coordinating the homeostatic responses to exercise. It has close links with the limbic system, which is much involved with emotions and behaviour, and with the basal ganglia where inhibition might make the fine control of motor output more difficult.

Although body temperature is an important limiting factor during endurance exercise, and may be critical during exercise in the heat, it is almost certainly not the only factor that can influence central drive. There is a multitude of afferent signals generated by exercise, ranging from sensations of increased ventilation to signals from joints, tendons and skin, and pain from blisters and inflamed tendons. With increasing duration of exercise, microtrauma to skin, joints and tendons leads to inflammation and increasing

inhibitory stimuli, which may impinge on the hypothalamus or other centres in the brain.

## CENTRAL STIMULANTS

Athletes, musicians, and all those needing to dance or work through the night, have always sought artificial ways of overcoming fatigue, frequently coming into conflict with the law, and sometimes risking their health in the process. As with any exciting but illegal area of activity, there is a paucity of objective information and a wealth of rumour and speculation. Nevertheless, this is an important area of study, both to understand the risks and benefits involved and also because anything that may modify fatigue will also give clues about the fundamental mechanisms involved.

The most commonly used substances include caffeine, amphetamine, cocaine and ecstasy, of which caffeine is by far the most common, being found in tea, coffee, cola and other sports/leisure drinks, as well as being sold in tablet form to offset fatigue. Although caffeine, amphetamine and cocaine are all believed to improve athletic performance, there is only a small literature involving human subjects concerning caffeine and, because of the obvious ethical problems, virtually nothing on amphetamine and cocaine.

### Caffeine

The effects of caffeine on endurance are marked, improving endurance time by approximately 20% when subjects exercise at a constant level (Fig. 11.12).

The improvement in endurance is reflected in the lower rating of perceived exertion (Fig. 11.12B) and it is interesting to note that, as with the effects of cool ambient conditions (see Fig. 11.9B), caffeine improves perception from the start of exercise rather than slowing the rate of development of the sensations of fatigue.

Caffeine is a remarkable substance, having a wide range of actions on different tissues, and consequently there are a number of ways in which it might act to improve performance. As discussed in Chapter 3, caffeine potentiates the release of calcium from the sarcoplasmic reticulum in muscle

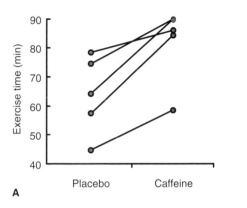

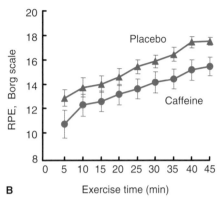

**Figure 11.12**  Effect of caffeine on endurance performance. **A**, exercise duration in five subjects taking placebo or caffeine ($5\,mg\,kg^{-1}$) when exercising at 75% $V_{O_{2max}}$. **B**, effect of caffeine on the rating of perceived exertion (RPE) during the first 45 min of endurance exercise, as in A. (Redrawn from the data of Bridge et al 2000.)

fibres, and it is possible that such an action would improve muscle performance and thus prolong endurance. Apart from the fact that fatigue during endurance exercise is largely of central origin, there are other reasons for thinking this unlikely. With isolated muscle preparations a minimum concentration of about $1\,mmol\,L^{-1}$ caffeine is required to affect contractile properties, but the circulating levels of caffeine in humans, even after a large dose, are an order of magnitude less than this. Caffeine is a potent inhibitor of the enzyme phosphodiesterase, increasing cellular levels of cyclic AMP, which stimulates both glycolysis and lipolysis. Increased fatty acid availability and oxidation could improve endurance (see above) and it is a common observation that caffeine increases circulating levels of free fatty acids. Interestingly, however, these increased plasma levels do not translate into increased oxidation, possibly another indicator that it is fatty acid uptake into the muscle, rather than availability, that is the limiting step in their metabolism. Having thrown doubt on peripheral mechanisms of action, the evidence suggests that caffeine acts centrally to modify perceptions of exertion and prolong exercise.

As temperature plays such an important role in limiting endurance, the first possibility is that caffeine may modify the hypothalamic pathways involved in thermoregulation and the neuroendocrine and behavioural responses described above. However, there is no evidence that this is the case: caffeine is equally effective with exercise at high and low ambient temperatures and does not modify the prolactin response to exercise. Caffeine is a methylxanthine and, having a structure similar to that of adenosine, it acts as an antagonist at $A_{2A}$ purinergic receptors. Adenosine, in turn, acts to inhibit the release of dopamine, so the net effect of caffeine is to potentiate the release of dopamine and thereby stimulate the activity of dopaminergic pathways.

## Dopamine and exercise

There are two major central neural pathways involving dopamine that may be important in exercise: one is the ventrolateral tract, terminating in the nucleus accumbens, and the other is the basal ganglia, especially the portion involving the substantia nigra.

The nucleus accumbens is of interest because of its involvement with addiction. The nucleus provides reward and reinforcing stimuli, which involve the release of dopamine. Cocaine and amphetamine are potent dopamine reuptake inhibitors (and possibly releasing factors) in the nucleus accumbens and consequently provide a powerful pleasure sensation, which can easily become addictive. Caffeine, which is mildly addictive, may have a similar action to cocaine and amphetamine in as much as it can increase dopamine release in the nucleus accumbens, albeit by a slightly different mechanism. Consequently one way in which these central stimulants could act is to provide a pleasurable response that masks the unpleasant stimuli arising during exercise (increased temperature, afferent information from pain receptors, etc.). Subjects who have taken caffeine are prepared to continue exercise and increase their body temperature beyond that which they would normally countenance. Although the action of ecstasy is not fully understood, it is a derivative of amphetamine and it is likely that at least some of the fatalities associated with its use are due to people becoming overheated having failed to respond to the usual protective reflex to high body temperature and stop dancing.

The second possible action of dopaminergic agents on exercise is via an effect on the basal ganglia. Probably the best known defect of the basal ganglia is Parkinson's disease, where there is degeneration of dopaminergic neurons in the substantia nigra. In this context it is interesting to note that fatigue is one of the complaints of these patients, although it is not the most serious of their problems. Clearly dopamine is essential to the normal function of the basal ganglia, and caffeine is known to interact with purinergic receptors in the striatum. Consequently it is possible that caffeine potentiates the function of the basal ganglia. The basal ganglia coordinate complex motor tasks and it may be that with prolonged activity some inhibition builds up so that coordinated, repetitive movements, such as running or cycling, become less automatic and require a greater conscious effort to carry out. Caffeine may reverse this process.

The neuropharmacology of exercise and fatigue is in its infancy and the discussion above is probably a simplistic and naive summary. There are almost certainly other pathways and transmitters involved, such as the opioids and cannaboloids, and more complex interactions between pathways than are even hinted at here.

## EFFECTS OF TRAINING ON CENTRAL ASPECTS OF FATIGUE

The one clear consequence of endurance training is to improve thermoregulation so that the rise in body temperature is at least slowed and, possibly, thermal equilibrium achieved, thereby reducing the inhibitory actions of high body temperature. The natural increase in body temperature as a result of exercise provides a good stimulus for adaptation but the response is more rapid, and possibly larger, if training is undertaken in a hot, humid environment.

Whether endurance training induces changes in the neural pathways discussed above remains to be seen. Experience suggests that training and familiarity are important factors allowing subjects to tolerate the various sensations associated with exercise such as breathlessness, heat, pain, etc., but to date there have been no satisfactory training studies in which it has been possible to demonstrate changes in the function of central pathways that may be involved.

REFERENCES AND FURTHER READING

Bergström J, Hultman E 1966 Muscle glycogen synthesis after exercise: an enhancing factor localized to muscle cells in man. Nature 210: 309–310

Bergström J, Hermansen L, Hultman E, Saltin B 1967 Diet, muscle glycogen and physical performance. Acta Physiologica Scandinavica 71: 140–150

Bloom S R, Johnson R H, Park D M, Rennie M J, Sulaiman W R 1976 Differences in the metabolic and hormonal response to exercise between racing cyclists and untrained individuals. Journal of Physiology 258: 1–18

Bridge M W, Broom J, Besford G, Allen A, Sharma A, Jones D A 2000 The action of caffeine and perception of exertion during prolonged exercise. Journal of Physiology 523: 224P

Brisson G R, Audet A, Ledoux M, Matton P, Pellerin-Massicotte J, Peronnet F 1986 Exercise-induced blood prolactin variations in trained adult males: a thermic stress more than an osmotic stress. Hormone Research 23: 200–206

Costill D L, Coyle E, Dalsky G, Evans W, Fink W, Hoopes D 1977 Effects of elevated plasma FFA and insulin on muscle glycogen during exercise. Journal of Applied Physiology 43: 695–699

Davis J M 1995 Central and peripheral factors in fatigue. Journal of Sports Science 13: S49–S53

Davis J M 1996 Nutritional influences on central mechanisms of fatigue involving serotonin. In: Maughan R J, Shirreffs S M (eds) Biochemistry of exercise IX, pp 445–455. Champaign, Illinois: Human Kinetics

Fink W J, Costill D L, Van Handel P J 1975 Leg muscle metabolism during exercise in the heat and cold. European Journal of Applied Physiology 34: 183–190

Gollnick P D, Karlsson J, Piehl K, Saltin B 1974 Selective glycogen depletion in skeletal muscle fibres of man following sustained contractions. Journal of Physiology 241: 59–67

Gonzalez-Alonso J, Clabet J A L, Nielsen B 1999a Metabolic and thermodynamic responses to dehydration-induced reductions in muscle blood flow in exercising humans. Journal of Physiology 520: 577–589

Gonzalez-Alonso J, Teller C, Andersen S L, Jensen F B, Hyldig T, Nielsen B 1999b Influence of body temperature on the development of fatigue during prolonged exercise in the heat. Journal of Applied Physiology 86: 1032–1039

Jeukendrup A E, Saris W H M, Wagenmakers A J M 1998a Fat metabolism during exercise: a review. Part I: Fatty acid mobilization and muscle metabolism. International Journal of Sports Medicine 19: 231–244

Jeukendrup A E, Saris W H M, Wagenmakers A J M 1998b Fat metabolism during exercise: a review. Part II: Regulation of metabolism and the effects of training. International Journal of Sports Medicine 20: 293–301

Kiens B, Essen-Gustavsson B, Christensen N J, Saltin B 1993 Skeletal muscle substrate utilisation during sustained endurance exercise in man: effect of endurance training. Journal of Physiology 469: 459–478

Marvin G, Sharma A, Aston W, Field C, Kendall M J, Jones D A 1997 The effects of buspirone on perceived exertion and time to fatigue in man. Experimental Physiology 82: 1057–1060

Newsholme E A, Blomstrand E 1996 Tryptophan, 5-hydroxytryptamine and a possible explanation for central fatigue. Advances in Experimental Medicine and Biology 384: 315–320

Nielsen B, Hales J R S, Strange S, Christensen N J, Warberg J, Saltin B 1993 Human circulatory and thermoregulatory adaptations with heat acclimation and exercise in a hot, dry environment. Journal of Physiology 460: 467–485

Pitsiladis Y P, Maughan R J 1999 The effects of exercise and diet manipulation on the capacity to perform prolonged exercise in the heat and in the cold in trained humans. Journal of Physiology 517: 919–930

Rennie M J, Winder W W, Holloszy O 1976 A sparing effect of increased plasma fatty acids on muscle and liver glycogen content in the exercising rat. Biochemical Journal 156: 647–655

Van Hall G, Raaymakers J S H, Saris W H M, Waggenmakers A J M 1995 Ingestion of branched-chain amino acids and tryptophan during sustained exercise in man: failure to affect performance. Journal of Physiology 486: 789–794

Wilson W M, Maughan R J 1992 Evidence for a possible role of 5-hydroxytryptamine in the genesis of fatigue in man: administration of paroxetine, a 5-HT re-uptake inhibitor, reduces the capacity to perform prolonged exercise. Experimental Physiology 77: 921–924

## QUESTIONS TO THINK ABOUT

1. Why does maximum running speed decline when muscle glycogen reserves are used up?

2. What limits the power output that can be achieved using fatty acids as a fuel for oxidative metabolism?

3. If fatty acid oxidation were to increase, what effect would this have on carbohydrate metabolism during exercise?

4. In what circumstances are peripheral mechanisms likely to predominate during endurance exercise?

5. What is meant by steady-state exercise and what factors may contribute to fatigue when such exercise is prolonged?

6. Describe the factors that will influence core body temperature during exercise.

7. What evidence is there that body temperature limits exercise capacity?

8. If body temperature acts to inhibit motor activity, what is the possible advantage of this reflex?

9. How may hypothalamic function (or at least one aspect of it) be monitored during exercise and how is this affected by body temperature?

10. What are the possible risks of taking central stimulants to prolong exercise?

# 12

# Muscle pain and damage

Muscular pain during and after exercise must be one of the most commonly experienced forms of discomfort, yet relatively little is known about its causes. There are two main types of pain associated with exercise, one arising during the activity and the other of delayed onset, occurring some time afterwards and generally called 'muscular stiffness' but often described in the literature as delayed-onset muscle soreness (DOMS) (Armstrong 1984). This chapter is concerned mainly with these two forms of muscular pain but will first deal briefly with the classification of sensory nerves and their endings. (See Further Reading for more on the subject of pain.)

## SENSORY NERVES

There are a variety of afferent sensory nerves in muscle, which are categorized according to size, conduction velocity and whether or not they are myelinated. The main types are listed in Table 12.1.

The small myelinated (type III or Aδ) and unmyelinated (type IV or C) fibres are concerned with signalling pain. These nerves branch freely in muscle and their unencapsulated endings are particularly dense in the myotendinous junctions and fascial sheaths. Stimulation of type III fibres gives rise to an immediate sensation of pain described as short lasting, well localized and of a pinprick quality, whereas simulation of type IV fibres results in more diffuse pain, of longer duration and with a dull or burning quality.

The type IV fibres have a complex function. Not only do they act as afferent sensory nerves but they also secrete substance P, a polypeptide that

**Table 12.1** Classification of sensory nerve fibres

| Afferent nerve | Type | | Receptor |
|---|---|---|---|
| Aα | Ia | } | |
| | Ib | } Large, myelinated | } Muscle spindles |
| Aβ | II | } | Tendon organs |
| Aδ | III | Small, myelinated | } |
| C | IV | Small, unmyelinated | } Nociceptors |

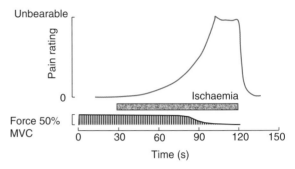

**Figure 12.1** Relationship between contractile activity, pain and ischaemia. The subject performed 50% maximal voluntary contractions (MVC) of the adductor pollicis, first with an intact circulation and then with the circulation occluded. Pain was estimated using an open-ended analogue scale.

causes vasodilatation and stimulates mast cells to degranulate and release histamine. Substance P also sensitizes afferent nerve endings, giving rise to the hypersensitivity to touch and heat that is associated with the region around an injury.

Most muscle pain receptors are thought to be polymodal, that is, they respond to mechanical, thermal or chemical stimuli, and one stimulus may lower the threshold for another. Agents that are known to activate afferent fibres include potassium, $H^+$, bradykinin, serotonin and histamine. Substances that sensitize nerve endings include substance P, prostaglandins and the leukotrienes.

## MUSCLE PAIN DURING ACTIVITY

### Ischaemic muscle pain

Discomfort develops rapidly in working muscles during high-intensity exercise such as running up a long flight of stairs. Although specific mechanisms have still to be identified, the general nature of the processes involved and the origin of the pain-producing substance can be inferred from a number of simple observations on muscle working with, and without, a circulation. Most physical exercise uses large muscle groups but the general mechanisms involved can be conveniently investigated using the small muscles of the hand and forearm.

Rhythmic submaximal voluntary contractions every 1–2 s can usually be sustained for long periods without pain, provided there is a normal circulation. However, pain develops rapidly if the circulation is occluded (often described as a burning sensation), reaching intolerable levels in a matter of minutes (Fig. 12.1).

When the contractions stop, the pain continues at much the same level until the circulation is restored and it then disappears within 10–20 s. These observations were first made by Sir Thomas Lewis, who concluded that, as a result of ischaemic exercise, some substance, probably a metabolite, was released from the muscle fibres and stimulated nociceptors. When the circulation was restored, the substance was washed away and the pain resolved. Lewis showed that the onset and intensity of the pain was related to the metabolic cost of the contraction: the higher the force and frequency, the more rapidly the pain developed. Short-term ischaemia of resting muscle does not produce painful sensations but if the muscle is kept without a circulation for 15–30 min, appreciable changes in muscle metabolites take place and the same painful sensations slowly develop.

The algesic substance has not been positively identified; potassium, inorganic phosphate and various nucleotides may all be released from fatiguing muscle and could, separately or together, activate pain fibres. Lactate and hydrogen ions have long been considered to be the cause of all muscle aches and pains but, as with the cause of muscle fatigue (see Chs 9 & 10), acidosis cannot be the only cause of pain and is unlikely even to be a major factor. Patients with myophosphorylase deficiency, who cannot produce lactate and whose muscles become slightly alkalotic during activity, develop pain rapidly during ischaemic contractions. While it is impossible to be certain that the

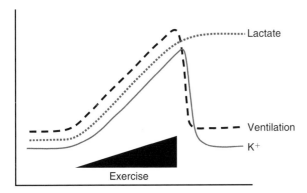

**Figure 12.2** Relationship between plasma lactate, potassium and ventilation during graded exercise. (Loosely based on data from Paterson et al 1990.)

sensations the patients experience are the same as those felt by normal subjects, the patients' descriptions of the pain, and reactions to it, are the same as those of anybody else. In addition, in normal subjects, at the end of exercise muscle remains acidic for 1–2 min after the circulation has been restored (see Fig. 10.19; Cady et al 1989), and after a bout of high-intensity exercise plasma lactate remains high for 5–10 min, long after the sensation of pain has gone from the muscle.

The most likely alternative to $H^+$ as the painful stimulus is potassium ions moving out of the muscle fibre during electrical activity, and the time course of the pain is similar to that of potassium movements. An instructive experiment was one in which $K^+$ and lactate were compared as stimuli for increased ventilation during exercise (Fig. 12.2). Plasma lactate and potassium levels increased in parallel with ventilation during the course of graded exercise, and either might have been the stimulus. However, at the end of exercise ventilation returned to resting values over the course of 30 s, or so, as did plasma potassium, whereas the lactate concentration remained high. The conclusion was that if either was a stimulus for ventilation it was more likely to be potassium than lactate. Similar arguments could be applied to the muscle pain experienced during high-intensity exercise.

The motive for the work on ischaemia in the 1930s and 1940s was a desire to understand the nature of the pain in patients with intermittent claudication, where the blood supply to a working limb is restricted by peripheral vascular disease, and to angina where coronary vessels are constricted. Lewis and colleagues were able to demonstrate that pathological pain had the same characteristics as pain produced in a normal muscle working under ischaemic conditions. Because the pain occurring during high-intensity exercise is the same as that caused by claudication or experimental ischaemia, it is often referred to as 'ischaemic pain' and will be so called in the following sections.

Patients with mitochondrial myopathies or deficiencies of the glycolytic pathway have larger metabolic changes in the muscle for a given amount of exercise and also develop pain in the working muscle more rapidly than normal subjects. Interestingly, hypothyroid patients, who have a low energy turnover when contracting their muscles, develop muscle pain relatively slowly, emphasizing the relationship between metabolic cost and pain during high-intensity exercise.

## Compartment syndrome

Compartment syndrome is a common and painful condition amongst athletes, which develops during exercise, most often in the anterior part of the lower leg (Hutchinson & Ireland 1994, Touliopolous & Hershman 1999). The pain, which is similar to ischaemic pain, develops with exercise more rapidly than would be expected and then persists longer than normal; in some respects it is as if the muscle were working and then remaining under ischaemic conditions. The anterior tibial compartment is bounded by bone on one side and a thick and relatively inextensible fascial sheath on the other. There is speculation that in susceptible subjects the increase in muscle volume, which occurs naturally with exercise as a result of increased blood flow and the osmotic effects of glycolysis, produces intramuscular pressures within the confined space of the compartment that effectively cut off the blood supply, rendering the muscle ischaemic and preventing, or at least seriously delaying, recovery. Whilst thickened fascia and raised intramuscular pressures have been found and in some cases fasciotomy is effective in relieving the pain, there are many conflicting reports and the cause of the condition is far from clear. It is likely that the symptoms may be the consequence of a number of different problems, including hypersensitivity to painful stimuli and, in some cases, undetected stress fractures of the tibia, as well as the circulatory problems that have received most attention.

## Cramp

The pain associated with cramp is distinctive and very distressing when it occurs in major muscle groups. It is associated with large involuntary contractions of the affected muscles and the pain may, in part, be similar to ischaemic pain. This cannot be the complete explanation, however, because the pain of a cramp is felt immediately the contraction begins, as opposed to the 15–30-s delay experienced with normal ischaemic contractions, and the quality is also different. There are usually some feelings of discomfort before the muscle goes fully into the contraction, and afterwards there is a feeling of tenderness not usually associated with even prolonged high-force isometric contractions.

Cramps can develop after prolonged exercise; extra time in football used to be the classical situation, although this does not seem to be so common in recent years. However, cramps can also develop after a single movement, such as back flexion during diving and in the calf muscles during the night. It is sometimes possible to produce cramp-like sensations by voluntarily contracting a muscle in a shortened position.

Cramps are electrically active contractions and characteristically develop only in muscles that are shortened. The cramp can be relieved by stretching the cramping muscle (however brutal this may seem at the time) and by stimulating the skin either with electricity, by rubbing, or with a cold spray (Schwellnus et al 1997). These characteristics suggest that the problem is one of increased excitability of the α motoneurons. Stretching the muscle stimulates Golgi tendon organs, which have inhibitory inputs to the α motoneurons. Skin afferents likewise have inhibitory inputs. Why α motoneurons should become hyperexcitable is

not known, but prolonged repetitive activity may facilitate neural pathways by a reduction of inhibitory input to the α motoneurons. In other situations, changes in the extracellular environment, such as increased potassium or decreased ionized calcium levels, may alter neuronal membrane excitability.

Increased sodium intake is usually recommended to prevent cramps, but there is no theoretical rationale for this and no evidence that it works. Quinine is also recommended and might act to reduce motoneuron membrane excitability but, again, there is no evidence either way to show that it is effective.

The mechanism of cramp is probably similar to that causing spasm of the back muscles where damage to vertebral joints sets up protective reflexes that make the α motoneurons very excitable. Spasm can be effectively managed by giving γ-aminobutyric acid (GABA) agonist drugs, which inhibit neuronal activity at both cortical and spinal level, and it would be interesting to know whether cramp can be managed in a similar way.

## Myalgic pain

Tenderness may develop in muscles, especially in the neck and shoulders, subsequently leading to prolonged pain, headaches and muscle fixation. This type of pain is experienced by a large proportion of the population and is of considerable economic importance because it affects people at work carrying out repetitive tasks such as typing and assembly work. Painful shoulder and neck muscles are also frequently seen following infection with Coxsackie B virus. Often areas can be identified from which the pain appears to radiate; these are known as 'trigger points'. Massage, heat and injection of local anaesthetics into these points are often felt to be of benefit. As with many types of muscle pain, there is speculation that the pain is due to local areas of muscle contraction, possibly as a reflex in response to tissue damage, generating ischaemic pain, but evidence for this cause is slight. Biopsies of the trigger points have shown no striking abnormalities and there is no evidence of increased electrical activity in the painful muscle while at rest. The primary lesion could well be in the sensory nerves serving the muscle which, having become sensitized by damage or a viral infection, transmit inappropriate information that is interpreted as pain.

# PAIN DEVELOPING AFTER EXERCISE

It is a common experience that heavy or unaccustomed muscular work is associated with a type of pain that becomes evident some hours after the end of exercise and may persist for several days. Typically it is experienced after the first match or training session of a season or when the game has been unusually hard. Discomfort develops in the muscle 6–12 h after the exercise and, typically, a subject will go to bed with only minor discomfort but wake the next morning with severe pain first appreciated while bending and trying to get out of bed. At rest there is little or no discomfort and it is not until the muscle is moved or touched that the pain is experienced. There are a number of sensations, of which the two most prominent are stiffness and tenderness, although in severe cases there is also a dull throbbing pain with the muscle at rest.

## Exercise causing delayed-onset pain

Asmussen (1953, 1956) was the first to document the fact that delayed-onset pain is caused mainly by movements in which the active muscle is stretched, such as whilst walking downhill or when lowering weights. This type of movement is often referred to as 'eccentric exercise', 'negative work' or, by some authors, as 'plyometric exercise'. Muscle groups commonly used to demonstrate this phenomenon are the quadriceps, stretched by stepping down from a box (Newham et al 1983a), the calf muscle, exercised by walking backwards down an inclined treadmill (Newham et al 1986), and the elbow flexors, stretched by forcibly extending the forearm (Jones et al 1986). In every case, delayed muscle pain develops only in the muscles that have been stretched. When stepping on and off a high box the quadriceps muscles of the leg used to step up shorten and perform work and, if the exercise is fast enough, the muscle fatigues and ischaemic pain is experienced which

resolves rapidly at the end of exercise. The quadriceps of the opposite leg, used to lower the body, is stretched whilst active and absorbs energy. The muscles do not become painful during the exercise and the only unusual sensation in the eccentrically exercised leg is of an ill-defined inability to control the muscle, evident as an increasingly heavy landing as the subject steps down. In experiments where the elbow flexors are forcibly extended, the subject is aware of very high forces in the muscle and tendons but there is no sensation that can be specifically defined as pain in the muscle. At the end of the exercise the subject often has increased tremor in the limb, and in the case of biceps it is difficult to bend the arm to touch the shoulder. The arm also hangs with the elbow in flexion (Jones et al 1987a).

Although delayed-onset muscle pain is most commonly associated with eccentric contractions, it can also be induced by high-force isometric contractions if the muscle is exercised at a long length (Jones et al 1989), showing that it is not entirely the high forces that result in muscle pain and that muscle length is an important factor.

## Muscular stiffness

This sensation is most evident 24–48 h after the exercise: difficulty and pain are experienced when stretching the affected muscles, for example when going down stairs or lowering the body on to a chair. There is a sensation of mechanical stiffness, which gives the feeling that the muscle has become shorter. After eccentric exercise of the elbow flexors there is difficulty straightening the elbow, and if it is straightened forcibly there is a very painful sensation in the belly of the muscle. With continuing activity the sensations diminish and after a few minutes of 'warm up', movements become fairly normal but, when the muscle is again rested, the sensations of stiffness return within 10–15 min with a similar intensity. When stiff, the muscle feels tense and swollen, and with the elbow flexors there is clear evidence of oedema, the swollen muscle increasing in circumference by 10–15% 1–2 days after exercise. A day or so later the fluid drains down and accumulates around the elbow.

Stiffness can be quantified by measuring the force required to straighten the elbow forcibly and is found to be maximal 1–2 days after the end of exercise, disappearing within about 5 days (Jones et al 1987a).

The stiffness and flexion contracture developed in the forearm flexors is electrically silent, which implies that the muscle fibres are not actively contracting or in spasm (Howell et al 1986, Jones et al 1987a). A possible explanation for the mechanical stiffness is increased fluid in the muscle as a result of oedema, but it is also possible that calcium is released from the sarcoplasmic reticulum of damaged fibres producing an electrically silent contracture.

## Muscle tenderness

Tenderness is an abnormally sharp sensation felt when moderate pressure is applied to the skin over a damaged muscle. While at rest and with no pressure on the muscle, there is no pain. The degree of tenderness can be measured by recording the threshold force required to elicit a painful response and this provides an objective measure of the extent and distribution of the pain (Edwards et al 1981). The feeling is similar to, if not identical with, that around a bruise or tendon sprain. Although it makes many actions unpleasant, for example sitting when the gluteal muscles are affected, there is no indication that tenderness inhibits the activation of muscles during isometric contractions, although there is likely to be considerable inhibition during contractions in which the muscle is stretched.

The tenderness, which in some cases can make a muscle acutely sensitive to touch, is maximal 1–2 days after eccentric exercise and is generally absent or much reduced by 5 days (see Fig. 12.5), having a similar time course to the muscle stiffness described above.

Muscle stiffness and tenderness are often ascribed to an accumulation of lactic acid in the affected muscle, but this is quite clearly not the case. First, following stepping exercise, the delayed-onset pain and stiffness is experienced in the leg that has done least metabolic work. Eccentric exercise entails minimal metabolic cost and it is most unlikely that there would be a significant accumulation of lactate. Second, the pain is apparent 24–48 h after exercise and any lactate that had accumulated during the exercise would have been metabolized or washed out of the muscle long before this.

## Inflammation and muscle tenderness

The cause of the pain and oedema is not known but it seems likely that it is an inflammatory response. Type IV (C) fibre endings secrete substance P, which stimulates mast cells to release histamine and small blood vessels to dilate. In addition, platelets coming into contact with damaged tissue release serotonin, and bradykinin will be formed from circulating kininogen.

Histamine, bradykinin and serotonin all cause pain. In addition, substance P from the type IV afferent fibres, and prostaglandins and leukotrienes from damaged tissues, sensitize afferent nerve endings (Fig. 12.3). Sensitization of pressure receptor nerve endings will cause what is normally perceived as touch to become a painful sensation, and this is the feature of tenderness that develops in a muscle that has been exercised in this way.

Tenderness and oedema are signs of classical inflammation and, as such, might be expected to be amenable to treatment with anti-inflammatory agents; however, aspirin and other anti-inflammatory agents have been found to have little or no effect on the muscle pain, suggesting that any sensitization is the result of substance P, rather than the action of prostaglandins or leukotrienes.

## Reflex activity

On the basis of EMG recordings from painful muscles, De Vries (1966) suggested that damage may result in increased reflex contraction of the muscle and consequent local areas of painful ischaemia. Other studies have produced no evidence to support this view, finding that painful muscles are electrically silent (Howell et al 1986, Jones et al 1987a). The suggestion also seems unlikely because the acute ischaemic pain experienced during a contraction is not the same sensation as tenderness.

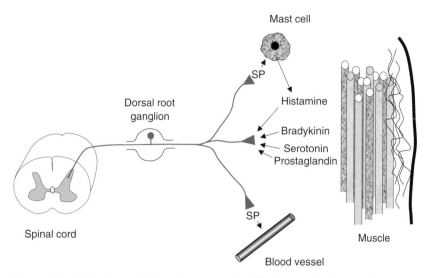

**Figure 12.3** Role of type IV fibres in muscle pain following muscle damage. Damage to muscle, including connective tissue and fascia, stimulates pain fibres. In addition, substance P (SP) released from nerve endings causes mast cells to degranulate and small blood vessels to dilate.

Anyone with severe DOMS will readily accept that the pain they were suffering was the result of muscle damage, and consequently we will now explore the nature of exercise-induced muscle damage to see how the pain and damage are related.

## MUSCLE PAIN AFTER ECCENTRIC EXERCISE IN RELATION TO DAMAGE

When box-stepping it is only the leg that supports the weight and is stretched during the downward phase that develops pain. The other leg, in which the muscles shortened while stepping up, does not become painful. The natural inference is that, while lengthening contractions are damaging, shortening or concentric contractions do no harm. There are several indications of damage, such as changes in contractile properties, that might be used to test this idea, but there can be confusion between what is fatigue and what is damage. The most clearcut evidence of damage is the release of soluble constituents from the muscle when the integrity of the fibre membrane breaks down. A number of substances could be measured, but myoglobin and creatine kinase (CK) are the most common and, in the case of CK, there is a specific skeletal muscle isoenzyme that can be measured should it be necessary to distinguish between damage to skeletal and cardiac muscle.

Figure 12.4 shows the changes in plasma CK levels following 45 min of exercise walking backwards down a gradient on a treadmill when the muscles were lengthening, compared with walking up the same gradient with the muscles shortening (Newham et al 1986). It was necessary to walk backwards downhill to ensure the muscles that were stretched were the same as those that shortened whilst walking uphill. Severe delayed-onset pain was experienced by all subjects only after walking downhill. The pain was experienced in the calf, gluteal and quadriceps muscles of both legs. With both uphill and downhill walking there was a suggestion of a small rise in plasma CK concentration 24 h after the exercise, but the most notable feature was the very large rise in the circulating CK level about 4–8 days after the eccentric exercise. In some subjects the rise was very large. Clinically, a value above about $1000\,\text{IU}\,\text{L}^{-1}$ is considered indicative of serious muscle disease, whereas in the experimental situation values were rising as much as ten times higher than this.

Although the size of the CK response to eccentric exercise is remarkable, the other interesting

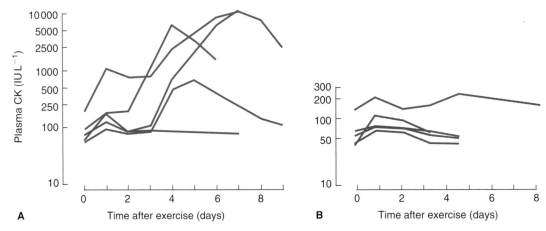

**Figure 12.4** Release of muscle enzymes following different types of exercise. **A**, after 45 min walking backwards down an elevated treadmill; **B**, after walking up the same gradient. Note that the scales for plasma creatine kinase (CK) concentration are both logarithmic. (Redrawn from the data of Newham et al 1986.)

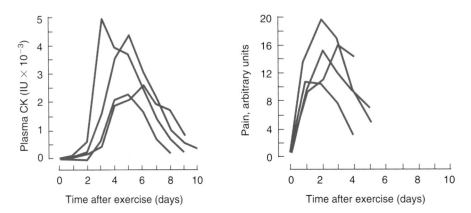

**Figure 12.5** Time course of muscle pain and plasma creatine kinase (CK) concentration following eccentric exercise of the forearm flexors. Subjects performed 80 maximal arm extensions and were then assessed at daily intervals. (Redrawn from the data of Jones et al 1986.)

feature is the delay of several days before the CK concentration begins to rise and the fact that it is maximal 4–7 days after the exercise. This delayed-onset muscle damage is reminiscent of delayed-onset muscle soreness, and it is interesting to compare the time course of the two phenomena.

Figure 12.5 shows the consequences of eccentric exercise of the forearm flexors (mainly the biceps) and it is evident that, although both pain and CK release are delayed after the exercise, the

time courses of the two are different. Pain peaked at 2 days, whereas the level of CK was only just beginning to rise at this time and did not peak until between 4 and 6 days, a time when the pain was absent, or at least much reduced.

The difference in time course of pain and CK release from the damaged muscle suggests that release of soluble constituents from damage to the muscle fibres may not be the cause of pain. However, CK is a relatively large molecule and

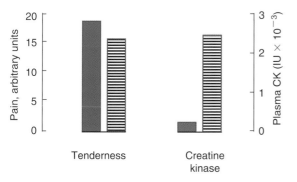

**Figure 12.6** Comparison of pain and creatine kinase (CK) response in two subjects who exercised by walking backwards down a gradient, as in Figure 12.5. Subject A (solid bars) had very painful muscles but no significant rise in CK levels, whereas subject B (horizontal lines) had both a large pain and CK response. (Redrawn from the data of Jones et al 1987b.)

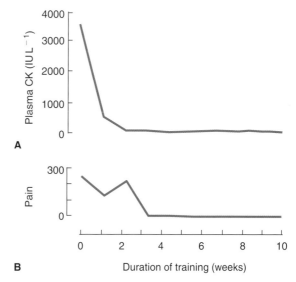

**Figure 12.7** Effects of weekly training on the release of creatine kinase (CK) from muscle (**A**) and on pain over the thigh (**B**). Stepping exercise was performed once every week for 10 weeks. (Redrawn from the data of Jones et al 1987b.)

it is possible that smaller proteins and peptides could be released at an earlier time when the muscle is painful, in which case a relationship might still be expected between the extent of muscle damage, as indicated by subsequent CK release, and the severity of the muscle pain.

The results in Figure 12.4 show there is considerable variation between subjects in the extent of the CK response, one subject having no significant increase despite very considerable tenderness. Figure 12.6 shows the results for two subjects in a similar experiment. Both subjects had similar levels of delayed-onset pain but only one had a large rise in plasma CK concentration.

It is always difficult to compare pain scores between subjects, because pain is so subjective, and subject A in Figure 12.6 might have been particularly sensitive to pain. However, it is possible to compare different levels of pain and damage within the same subject.

One of the features of eccentric exercise is the rapidity with which the muscle adapts to repeated exercise. Quadriceps tenderness and plasma CK concentration are shown in Figure 12.7 for a subject who performed a stepping exercise at weekly intervals. During the first week there was a large enzyme release, but in the following weeks no significant release occurred. In the first and second training weeks, therefore, there was appreciable pain but no evidence of muscle fibre damage in terms of CK release.

The subject used her right leg to step up and her left leg to step down throughout the training period. It is interesting to note that after 11 weeks the subject 'changed legs', stepping down with her right leg which had previously been used only to step up. In the days that followed the pain experienced in the right leg was as severe as that in her left leg had been on the first occasion 12 weeks before, and there was likewise a large increase in plasma CK concentration, showing that the adaptation was specific to the leg that had been trained.

Delayed-onset muscle pain and severe muscle fibre damage can both be consequences of eccentric muscle contractions, but the two events are separated in time and pain can be apparent in the absence of obvious muscle fibre damage. In this context it is interesting that muscle pain is not a feature of diseases such as muscular dystrophy, where there is severe damage to muscle fibres and high circulating levels of CK. An alternative explanation to muscle fibre damage as the cause of delayed-onset pain is that it is a consequence of damage to some other element of the muscle tissue. Muscle fibres are embedded in a matrix of collagen fibres and connective tissue, which form the tendons at the end of the muscle. Sprains and bruises are largely due to damage to connective tissue and are tender in much the same way as delayed-onset muscle pain. The time course of pain resulting from a sprain or bruise is also similar to that of muscle tenderness following eccentric exercise. It seems likely, therefore, that, in addition to any damage to muscle fibres, unaccustomed eccentric exercise damages connective tissue within the muscle. The release of serotonin from platelets, of histamine from mast cells and of substance P from C-fibre nerve endings as a result of connective tissue damage may all contribute to an inflammatory response, sensitizing pressure receptors and resulting in the characteristic sensation of tenderness.

## MUSCLE FIBRE DAMAGE FOLLOWING ECCENTRIC EXERCISE

There are a number of ways in which muscle damage may become apparent. The release of soluble components from disrupted surface membranes has already been mentioned, but there are also changes in strength and contractile properties of the muscle together with the structure of the muscle as revealed by conventional histology and EM.

Following eccentric exercise there is a reduction in the maximum force developed by the muscle but no effect on the speed as assessed by the rate of relaxation, showing that the changes differ from those seen in acutely fatigued muscle. One feature of the contractile properties that does change is the shape of the force–frequency relationship, shifting to the right due to a specific loss of force at the lower frequencies of stimulation (Jones 1996). It is convenient to report changes in the force–frequency relationship in terms of a ratio of forces developed at 20 and 50 Hz. Figure 12.8 shows changes in maximum force, the 20/50 ratio, muscle tenderness and the appearance of CK in the plasma in the days after eccentric exercise. Maximum force and 20/50 ratio fell to their lowest values immediately after the exercise, partially recovering over the next 10 days. In

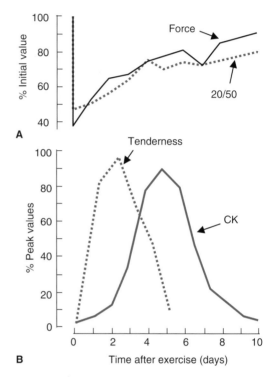

**Figure 12.8**   Changes in muscle force, contractile properties, pain and release of creatine kinase (CK) following eccentric exercise of the forearm flexors. **A**, maximum voluntary contraction (MVC) force and 20/50 ratio expressed as a percentage of pre-exercise values. **B**, pain and CK expressed as a percentage of peak values. (Redrawn from the data of Jones et al 1986.)

contrast, both the pain and CK release showed a characteristic delayed response, pain peaking at about 2 days and CK at 5 days. CK release from muscle is an indication of loss of membrane integrity and this is confirmed by measuring the uptake of a radioactive marker into the muscle, the extent of uptake having the same time course as CK release. This technique is also useful in identifying damaged muscles, or portions of muscles (Newham et al 1987). It is notable that the maximum destruction of fibres, as indicated by CK release, appeared to occur 5 days after the maximum force loss at a time when force was approximately 50% recovered, suggesting that the loss of cellular integrity is not the primary cause of the loss of muscle force.

As this form of muscle damage occurs when the muscle is stretched or exercised at a long length,

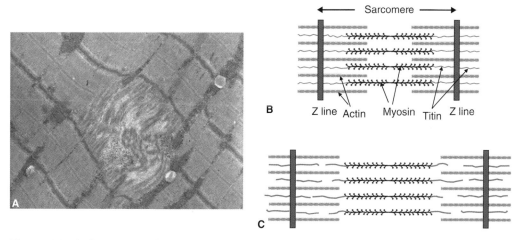

**Figure 12.9** A–C, disruption of sarcomere structure by eccentric exercise.

it has been suggested that the damage may be due to some sarcomeres becoming overextended and possibly breaking the titin filaments that help to maintain sarcomere structure (Fig. 12.9B,C) (Jones et al 1989, Morgan 1990). This has become known as the 'popping sarcomere' theory of muscle damage. It is easy to see why the damaged sarcomere in Figure 12.9A will not generate effective force, and it is likely that it will also interfere with the function of the intact sarcomeres, reducing the transmission of force along the length of the fibre.

The reason for the decrease in the 20/50 ratio is not clear. It could be due to damage caused to the T tubules and sarcoplasmic reticulum, so that less calcium is released per electrical impulse, or, alternatively, the disrupted sarcomeres might add series compliance thereby decreasing the relative size of the twitch (Jones 1996).

The recovery of function over days may be due to repair of the sarcomere damage shown in Figure 12.9. If this is the case, it raises the question of the significance of CK release and the damage that this represents. Light microscopy (Fig. 12.10) shows that 4 days after exercise, when function is beginning to return and the CK level is rising, the fibres appear relatively normal; at 7 days, when CK levels are beginning to decline, there is clear evidence of disrupted fibres and infiltration of the tissue by monocytes (mainly macrophages); and

at 12 days, when CK levels have returned to normal, there is maximum disruption of muscle fibres and infiltration of the tissue. Full recovery of structure and function occurs by approximately 20 days.

The damage to the calf muscle illustrated in Figure 12.10 is probably the most severe that can be produced in human muscle by voluntary exercise and it graphically illustrates one of the unexplained features of this type of damage. During the 4–12 days following the exercise, muscle strength and contractile properties were returning towards normal, albeit slowly. Despite this, the appearance of the muscle indicates increasing amounts of damage. One explanation of this discrepancy is that, while most, or all, fibres are damaged to some extent, there is a population that is more severely damaged and can generate no force following the eccentric exercise. As the less severely damaged fibres recover their function, the severely damaged fibres degenerate and are removed. Consequently the dramatic histological changes seen in Figure 12.10 between days 4 and 12 represent the removal of fibres that were making no substantial contribution to force and function. In the 2–3 days after exercise, these fibres lose their membrane integrity and release soluble components, including CK. Monocytes are then attracted into the tissue and, as macrophages,

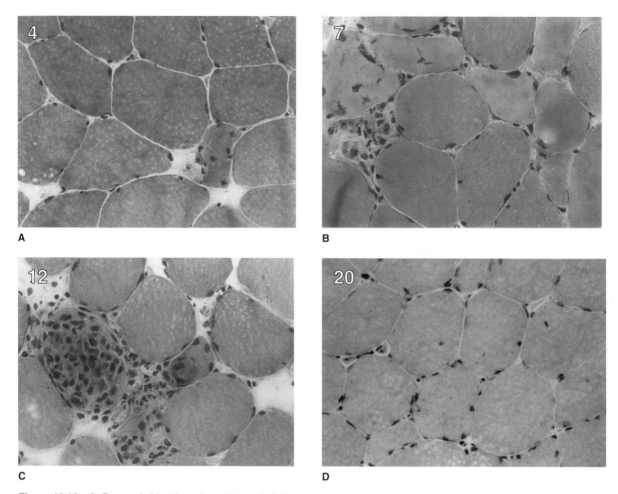

**Figure 12.10** **A–D**, muscle biopsies taken at intervals following eccentric exercise. Calf muscle damaged by downhill walking and biopsies taken 4, 7, 12 and 20 days later, as indicated.

remove cellular debris and probably stimulate satellite cells to divide and begin the regeneration of the muscle. This regeneration is a remarkably rapid process, as only 8 days separates the two biopsies in the lower row of Figure 12.10.

The nature of the suggested population of susceptible fibres is a matter of considerable speculation. It has been suggested that they are old fibres, and certainly muscles from aged animals are more susceptible to this type of damage than younger muscles. However, the concept of an old fibre is difficult to comprehend because

muscle fibres are not single cells but multinucleate aggregates. As individual nuclei are turning over all the time, it would be difficult to define the age of a whole fibre or imagine that there will be much difference between different fibres in the muscle. The most severe damage seems to affect the fast type 2 fibres (Jones et al 1986, Lieber & Fridén 1988), although why this should be is not known. Possibly it has to do with ultrastructural differences between fibre types, such as the thinner Z lines in the type 2 fibres (see Fig. 11.1).

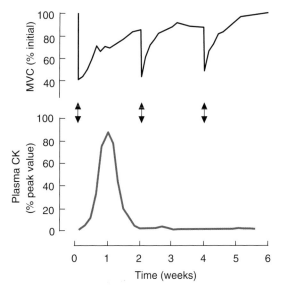

**Figure 12.11** Effect of repeated bouts of eccentric exercise on muscle force and creatine kinase (CK) release. Subjects performed eccentric exercise of the forearm flexors at 2-week intervals, indicated by the broad vertical arrows. MVC, maximum voluntary contraction.

## EFFECT OF TRAINING ON MUSCLE DAMAGE CAUSED BY ECCENTRIC EXERCISE

If subjects are trained by repeated bouts of eccentric exercise there is a very marked change in the amount of muscle damage estimated by the appearance of CK in the plasma (Fig. 12.11).

The protection afforded by a single bout of eccentric exercise lasts for between 6 and 12 weeks, but the mechanism responsible for the increased resistance to damage is not known. Whilst the trained muscle does not undergo the dramatic changes in structure and release of CK, it is notable that repeated exercise does not protect against the initial loss of force (Fig. 12.11), and the same is true of changes in the 20/50 ratio. Careful comparison of the rates of recovery of force following the first bout of exercise compared with the second and third occasions reveals a slower time course on the first occasion with a pause at about 5–7 days, the time when CK peaks (Fig. 12.11). This pause, and the release of CK, is absent in the subsequent recovery curves. The pause in recovery of force argues against the previous suggestions of a population of fibres that contribute nothing to force immediately after the exercise and are subsequently removed. Rather it appears that there are two concurrent processes going on in the damaged muscle, one of rapid repair, the other of degeneration. Whether or not a fibre survives may depend on the extent of the initial damage and the balance between the processes of repair and degeneration.

What changes occur in muscle to make it less susceptible to damage by eccentric exercise are not known, but they must be specific to the exercised muscle and are thus unlikely to be alterations in circulating hormone levels or of the immune system. One likely change is in the number of sarcomeres in series along the length of the muscle fibre. Passive stretch is known to be an effective stimulus for growth in muscle fibre length and sarcomere number, and it is likely that active stretching will have the same effect. More sarcomeres in series will mean that, for a given overall extension of the muscle, each sarcomere will be stretched by a smaller amount and so be protected against overextension and damage, as suggested in Figure 12.9C.

What changes protect against muscle pain are similarly not known, but an increase in connective tissue length or strength may make the muscle less susceptible to damage by subsequent eccentric exercise.

## OTHER MECHANISMS OF MUSCLE DAMAGE

### Physical trauma

By the nature of their position in the body, skeletal muscles are frequently subjected to physical trauma and this risk is increased by participation in contact sports. There is little information about the extent of muscle fibre damage, which occurs as a result of the bumps and bruises of everyday and sporting life, but probably most of the painful sensations are due to inflammation of the skin, subcutaneous tissue and muscle fascia, with relatively little involvement of the muscle fibres. Fibres are cut as a result of a muscle needle biopsy, but this procedure has remarkably little effect on muscle function and many of the classical studies of human muscle physiology have involved repeated biopsies during the course of prolonged activity such as 50-km ski races, with no apparent detrimental effect on the subjects or their performance. Muscle biopsy does not produce a significant rise in circulating CK levels. Orthopaedic surgery in which large amounts of muscle may be cut produces relatively modest rises in plasma CK concentration, and only occasionally are values seen that are comparable to those generated by experimental muscle damage, and the release is rapid, in the first day or so, after surgery (Jones et al 1991) rather than the delayed response seen with eccentric exercise.

## Crush and reperfusion injury

Muscles may become ischaemic as a result of an accident where some heavy object falls across a limb, an unconscious person lies so that blood is cut off or when the artery supplying the limb is damaged. Limb ischaemia is also used routinely during orthopaedic and vascular surgery. If the ischaemia is prolonged for more than 4–5 h, major damage occurs to the muscle when oxygenated blood returns to the tissue. Initially the limb seems to recover and becomes pink and apparently healthy, but oedema develops within hours and there is massive breakdown of muscle, known as rhabdomyolysis. This may be so severe that the myoglobin released from the damaged muscle precipitates in the kidneys, blocking the tubules and causing renal failure, which then requires dialysis for about 10 days until the kidney damage is repaired. Myoglobin and CK reach exceptionally high levels in the plasma and peak about 24 h after the start of reperfusion, again in contrast to eccentric exercise where there is a long delay. Muscle biopsy confirms the extent and time course of the muscle fibre damage (Adiseshiah et al 1992).

It is the return of oxygenated blood that precipitates the crisis. Muscle is relatively tolerant of this type of treatment, whereas other tissues such as gut and brain are very sensitive to just a few minutes of ischaemia followed by reperfusion.

The injury is probably a result of free radicals generated by xanthine oxidase, and the reason that muscle is relatively resistant may be because it has a low activity of this enzyme.

## Intracellular calcium and muscle damage

Intracellular calcium is maintained at very low levels, so there is approximately a 10 000-fold gradient between the interior and exterior of a muscle fibre. The extremely low intracellular concentration provides precise control of muscle contractile activity but it also regulates other functions of calcium, which include activation of various proteases and phospholipases involved in tissue degradation. With isolated muscle preparations it can be shown that metabolic depletion leads to an influx of calcium and subsequent muscle damage, and this can be prevented by removing external calcium (Jackson et al 1984). Prolonged activity entails a persistently raised intracellular level of calcium as it is released from the sarcoplasmic reticulum. In the long term this could be a signal for the modification of gene expression but, in the short term, it can lead to tissue damage. This latter process has been suggested as a mechanism of long-lasting low-frequency fatigue (Westerblad et al 2000), with calcium-activated enzymes degrading the excitation–contraction coupling mechanism. Calcium also enters muscle fibres via voltage-activated sodium channels in the surface membrane during the course of normal electrical activity and this can lead to damage by the same mechanism (Gissel & Clausen 2001).

## Muscle damage during prolonged activity

Most of the experimental work on muscle damage using human subjects has concerned eccentric exercise, and this is very relevant to sporting activities involving stretching. However, marathon runners and other endurance athletes may feel a little neglected as they also suffer from severe muscle stiffness and pain that has similar properties to the delayed-onset pain discussed above but do not experience the same extent of muscle stretch as occurs in other sports. Some of their pain and damage may, indeed, be as a result of the repeated small stretches that occur as the leading leg hits the ground and absorbs energy with every stride, and the pain is reported to be greater after marathons that have a substantial downhill component. However, this cannot be the full explanation because the pattern of CK release differs from that seen after eccentric exercise with peak levels being seen immediately after the marathon, or certainly within the next 24 h (Noakes 1987). The

cause of this type of damage is unknown but could be any one, or a combination, of the mechanisms discussed above. High levels of oxidative activity also generate free radicals (Duthrie et al 1996, Jackson 1996) that could cause damage.

Of the various possibilities, the only substantial evidence available, gathered after a 100 km race, favours an effect mediated by increased intracellular calcium levels (Overgaard et al 2002). How muscle fibre damage relates to the pain in this situation remains to be determined.

REFERENCES AND FURTHER READING

Adiseshiah M, Round J M, Jones D A 1992 Reperfusion injury in skeletal muscle: a prospective study in patients with acute limb ischaemia and claudicants treated with revascularization. British Journal of Surgery 79: 1026–1029

Armstrong R B 1984 Mechanisms of exercise induced delayed onset muscular soreness: a brief review. Medicine and Science in Sports and Exercise 18: 529–536

Asmussen E 1953 Positive and negative muscular work. Acta Physiologica Scandinavica 28: 364–382

Asmussen E 1956 Observations on experimental muscle soreness. Acta Rheumatologica Scandinavica 2: 109–116

Cady E B, Jones D A, Lynn J, Newham D J 1989 Changes in force and intracellular metabolites during fatigue of human skeletal muscle. Journal of Physiology 418: 311–325

De Vries H A 1966 Quantitative electromyographic investigation of the spasm theory of muscle pain. American Journal of Physiological Medicine 45: 119–134

Duthrie G G, Jenkinson A McE, Morrice P C, Arthur J R 1996 Antioxidant adaptations to exercise. In: Maughan R J, Shirreffs S M (eds) Biochemistry of exercise IX, pp 465–470. Champaign, Illinois: Human Kinetics

Edwards R H T, Mills K R, Newham D J 1981 Measurement of severity and distribution of experimental muscle tenderness. Journal of Physiology 317: 1–2P

Gissel H, Clausen T 2001 Excitation-induced $Ca^{2+}$ influx and skeletal muscle cell damage. Acta Physiologica Scandinavica 171: 327–334

Howell J H, Chila A G, Ford G, David D, Gates T 1986 An electromyographic study of elbow motion during postexercise muscle soreness. Journal of Applied Physiology 58: 1713–1718

Hutchinson M R, Ireland M L 1994 Common compartment syndromes in athletes: treatment and rehabilitation. Sports Medicine 17: 200–208

Jackson M J 1996 Free radicals, exercise and health. In: Maughan R J, Shirreffs S M (eds) Biochemistry of exercise IX, pp 459–464. Champaign, Illinois: Human Kinetics

Jackson M J, Jones D A, Edwards R H T 1984 Experimental skeletal muscle damage: the nature of the calcium-activated degenerative process. European Journal of Clinical Investigation 14: 369–374

Jones D A 1996 High and low frequency fatigue revisited. Acta Physiologica Scandinavica 156: 265–270

Jones D A, Newham D J, Round J M, Tolfree S E J 1986 Experimental human muscle damage: morphological changes in relation to other indices of damage. Journal of Physiology 375: 435–448

Jones D A, Newham D J, Clarkson P M 1987a Skeletal muscle stiffness and pain following eccentric exercise of the elbow flexors. Pain 30: 233–242

Jones D A, Newham D J, Obletter G, Giamberadino M A 1987b Nature of exercise-induced muscle pain. In: Tiengo M, Eccles J, Cuello A C, Ottoson D (eds) Advances in Pain Research and Therapy, vol. 10, pp 207–218. New York: Raven Press

Jones D A, Newham D J, Torgan C 1989 Mechanical influences on long-lasting human muscle fatigue and delayed onset pain. Journal of Physiology 412: 415–427

Jones D A, Round J M, Carli F 1991 Plasma creatine kinase in patients following routine surgery: a comparison with experimental muscle damage. Journal of Physiology 438, 173P

Lewis T 1942 Pain. New York: Macmillan

Lewis T, Pickering G W, Rothschild P 1931 Observations on muscular pain in intermittent claudication. Heart 15: 359–383

Lieber R L, Fridén J 1988 Selective damage of fast glycolytic muscle fibres with eccentric contraction of the rabbit tibialis anterior. Acta Physiologica Scandinavica 133: 587–588

Morgan D L 1990 New insights into the behaviour of muscle during active lengthening. Biophysical Journal 57: 209–221

Newham D J, Mills K R 1999 Muscles, tendons and ligaments. In: Wall P D, Melzack R (eds) Textbook of pain, pp 517–538. Edinburgh: Churchill Livingstone

Newham D J, Jones D A, Edwards R H T 1983a Large delayed plasma creatine kinase changes after stepping exercise. Muscle and Nerve 6: 380–385

Newham D J, Mills K R, Quigley B M, Edwards R H T 1983b Pain and fatigue following concentric and eccentric contractions. Clinical Science 64: 55–62

Newham D J, Jones D A, Edwards R H T 1986. Plasma creatine kinase after concentric and eccentric contractions. Muscle and Nerve 9: 59–63

Newham D J, Jones D A, Clarkson P M 1987 Repeated high-force eccentric exercise: effects on muscle pain and damage. Journal of Applied Physiology 63: 1381–1386

Noakes T D 1987 Effect of exercise on serum enzyme activities in humans. Sports Medicine 4: 245–267

Overgaard K, Lindstrom T, Ingemann-Hansen T, Clausen T 2002 Membrane leakage and increased content of $Na^+$–$K^+$ pumps and $Ca^{2+}$ in human muscle after a 100-km run. Journal of Applied Physiology 92: 1891–1898

Paterson D J, Friedland J S, Bascom D A et al 1990 Changes in arterial $K^+$ and ventilation during exercise in normal subjects and subjects with McArdle's syndrome. Journal of Physiology 429: 339–348

Schwellnus M P, Derman E W, Noakes T D 1997 Aetiology of skeletal muscle 'cramps' during exercise: a novel hypothesis. Journal of Sports Science 15: 277–285

Touliopolous S, Hershman E B 1999 Lower leg pain. Diagnosis and treatment of compartment syndromes and other pain syndromes of the leg. Sports Medicine 27: 193–204

Westerblad H, Bruton J D, Allen D G, Lannergren J 2000 Functional significance of $Ca^{2+}$ in long-lasting fatigue of skeletal muscle. European Journal of Applied Physiology 83: 166–174

## QUESTIONS TO THINK ABOUT

1. What are the different types of pain fibre and what function do they have?

2. In the type of experiment undertaken by Thomas Lewis on ischaemic pain, what happens to the blood pressure and what does this tell us about the role of type 3 and type 4 muscle afferents?

3. Describe how an inflammatory reaction occurs in response to tissue injury.

4. To what extent does muscle pain indicate muscle damage?

5. To what extent does muscle pain indicate a metabolic crisis in the muscle?

6. What are the best indicators of muscle damage?

7. Why is creatine kinase used so often as an indicator of muscle damage?

8. The training effect of eccentric exercise, reducing the amount of damage on the second and subsequent occasions, might be due to increasing the numbers of sarcomeres in series. How might this help to prevent damage and how could this idea be tested?

9. Delayed-onset muscle soreness (DOMS) can be temporarily relieved by gentle movements of the affected limb. What does this tell us about the nature of the pain?

10. What are the similarities and what are the differences between the pain and damage caused by eccentric contractions and by running a marathon?

# 13

# Muscle energetics and efficiency during exercise

Knowing the power, speed and maximum oxygen uptake of an athlete usually gives a fair idea of their potential but, at the highest levels, there is probably no significant difference in $VO_{2max}$ between the top 100 or even 1000 marathon runners in the world, and the question of why some of these runners consistently perform better than others has to be asked. There are many possible explanations, including mental attitude, tactical awareness and differences in thermoregulation, but one that has long been debated is a possible difference in their efficiency. Clearly, in a sport where the energetic demands are near to the limits of the body, small differences in the efficiency of converting chemical energy into mechanical work will allow faster running speeds, lower heat generation or better conservation of fuel reserves, all of which give a competitive edge.

Efficiency of whole body exercise (sometimes known as *gross efficiency*) is calculated from the measured work output and the oxygen cost of the exercise. There are a number of practical problems involved in these measurements (Cavanagh & Kram 1985). For running, it is difficult to quantify the work output because a significant proportion of the energy is required to lift the centre of gravity up and down with each stride. Another limitation is the fact that exercise must be in a steady state so that the oxygen consumption truly reflects the metabolic demands of the working muscles. It is also important that the proportions of fat and carbohydrate are known, as the oxygen costs of metabolizing these are different. Oxygen consumption ceases to be a true reflection of the metabolic costs of working muscles if there is an anaerobic component to the exercise, and the proportions of fat and carbohydrate cannot be calculated once the respiratory exchange ratio is greater than 1.

Despite the problems, the question of efficiency has been widely investigated with, for instance, differences in fibre type composition and cycling cadence being reported to make significant differences to efficiency and performance (Horowitz et al 1994). With cycling, an increase in efficiency as small as the order of the coefficient of variation of the measurement is estimated to improve performance by up to 1 min over a 36-km time trial—a major competitive gain.

Clearly, many factors affect the gross efficiency of an activity, including biomechanical considerations and the coordination of a subject, but the factor that is most amenable to experimental investigation is the efficiency of the working muscle itself, considering different types and speeds of contraction and the effect of fibre-type composition.

## ENERGY UTILIZATION IN MUSCLE

Although almost every cellular process requires ATP, in working muscle two processes use the majority of the available energy: the maintenance of ionic gradients, including that of calcium across the sarcoplasmic reticulum (SR), and the workings of the myosin cross-bridges in the generation of force and movement (Fig. 13.1).

With each action potential there is a movement of $Na^+$ into the muscle and of $K^+$ out, and if the muscle is to remain excitable these movements have to be reversed by the $Na^+$, $K^+$ transport ATPase in the surface membrane. The extent of ionic movements during activity will depend on the frequency of stimulation and the surface area of the muscle membrane. The major part of the surface area of a muscle fibre consists of the T tubular membrane, and this is more highly developed in fast fibres. Fast fibres are also activated at higher frequencies than slow fibres so the total ionic flux will be proportionately higher, as will be the energetic cost of restoring and maintaining the gradients of sodium and potassium across the surface membrane. In terms of the energy costs of

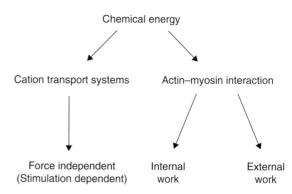

**Figure 13.1** Processes using chemical energy during muscle activity.

a contracting muscle, those attributable to $Na^+$, $K^+$-ATPase are thought to be of the order of only 3% of the total.

The release of calcium may be sensitive to the ATP levels in the fibre, but it is not believed actually to consume significant amounts of ATP. However, the reuptake of calcium from the cytosol into the SR is an important and energetically demanding process. It is important to realize that as soon as calcium is released it starts to be taken up by the $Ca^{2+}$ transport ATPase of the SR, so there is a continual flux of calcium throughout the contraction and not simply at the end. Distinguishing the energy requirements of the SR $Ca^{2+}$-ATPase from those of the myosin ATPase is not easy, but estimates can be made by looking at the effects of sarcomere length on force and energy consumption. On the right-hand side of the length–force relationship there is reduced cross-bridge interaction as muscle length increases (see Fig. 2.2). Consequently the ATP splitting associated with cross-bridge activity will decrease but that associated with calcium reuptake will remain constant, and thus extrapolating ATP splitting to zero filament overlap gives the energy requirement of the SR $Ca^{2+}$-ATPase. It is estimated that between 30% and 40% of the total energy requirement of a muscle contracting isometrically at optimum length can be attributed to SR function (Homsher & Kean 1978). Because fast fibres have highly developed SR systems and are activated at relatively high frequencies, the energy cost of calcium pumping will be high in these fibres compared with slow fibres.

However, although the absolute energy costs of handling intracellular calcium may be higher in fast fibres, the relative contribution is fairly similar in fast and slow fibres because the costs of cross-bridge activity is also high in fast fibres.

The relative contribution of the fixed cost of calcium handling will also vary with the type of contraction, being high when muscles are working at long lengths or during stretch and low when muscles are shortening and cross-bridges are turning over rapidly.

## ECONOMY AND EFFICIENCY

During an isometric contraction no external work is done and, strictly speaking, the efficiency is zero no matter what force is generated or however much ATP is hydrolysed. However, isometric contractions are important elements of daily activity and a useful concept is that of the *economy* of an isometric contraction. This refers to the energy cost of a contraction, related to the integral of force and time, otherwise known as the *impulse*. The economy of an isometric contraction is the energy cost per unit muscle impulse and it is important to note the difference between this and the *efficiency* of a shortening contraction, which is the energy cost per unit of *work* (Fig. 13.2).

## Muscle dimensions

The isometric force of a muscle is a function of the physiological cross-sectional area and is not dependent on its length; more cross-bridges in parallel mean more force, but there will be a corresponding increase in ATP turnover. In contrast, if the muscle cross-sectional area remains constant while the length increases, then the isometric force will remain the same but the number of cross-bridges will increase, as will ATP turnover, and the muscle economy will consequently decrease. During growth, muscle volume increases as a result of both increases in length and cross-sectional area. Force increases are related to increases in cross-sectional area alone, whilst ATP splitting is proportional to muscle volume, a function of both cross-sectional area and length. Consequently the economy of a muscle decreases during

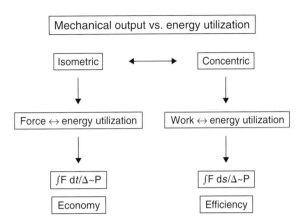

**Figure 13.2** Calculations of economy (isometric contractions) and efficiency (concentric contractions). In the former, force (*F*) is integrated with respect to time (d*t*), while in the latter it is with respect to distance (d*s*) and expressed per unit of high-energy phosphate used ($\Delta\sim$P).

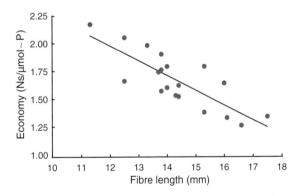

**Figure 13.3** Influence of fibre length on the economy of isometric contractions of medial gastrocnemius muscles of rats at various ages. The economy is calculated as impulse (force (N) $\times$ time (s)) divided by the ATP cost ($\sim$P) of the contraction. (Redrawn from the data of de Haan et al 1988.)

development, decreasing in proportion to the increase in length (Fig. 13.3).

Work is the force generated by a muscle multiplied by the distance moved. During shortening, each cross-bridge performs a small amount of work with every cycle of attachment, movement and detachment. The total work done is the sum of all the individual cross-bridge cycles. Consequently, to achieve a given total work it is possible to have a few cross-bridges turning over many times or a lot of cross-bridges turning over a few times. For a short fat muscle with a large number of cross-bridges in parallel it is necessary to activate a relatively small proportion to generate a given force, but these have to turn over a large number of times to achieve the required shortening. In contrast, for a long thin muscle it is necessary to activate a large proportion of cross-bridges to generate the force but, because there are so many cross-bridges in series, each cross-bridge has to turnover relatively few times for the whole muscle to shorten the required distance. If each cross-bridge cycle consumes one molecule of ATP, the total ATP turnover will be the same for the long and short muscle to perform the same amount of work. In other words, whilst the *economy* of isometric contractions differs between long and short muscles, the *efficiency* of shortening contractions is the same. As a muscle increases in volume, as a result of either growth during childhood or strength training as an adult, the efficiency should remain constant.

## Fibre composition

Type 1 fibres have lower rates of ATP turnover during isometric contractions than the faster type 2 fibres and, because the forces generated per unit cross-sectional area are similar, the overall economy of the two fibre types is different. In isolated intact animal muscles the difference in economy is reported to be about twofold (Crow & Kushmerick 1982, Goldspink 1978), whereas for skinned segments of human muscle the difference is found to be fourfold (Szentesi et al 2001). The greater economy of the type 1 fibres has obvious advantages as these fibres are used predominantly in prolonged, mainly isometric, contractions required to maintain posture, and they do so with minimal energetic demands. The reason for the differences between fibres arises from the nature of the myosin cross-bridges and their rate of cycling. Type 2 myosin goes through the cycle of attachment, rotation and detachment faster than the slow myosin and consequently, in a given time, there is a greater ATP turnover. In the isometric state, force production depends on the ratio of the rates of attachment and detachment, not the absolute values (see Ch. 2), and low

absolute values for the rate constants equate with low turnover and high economy. During shortening, however, the force that can be sustained depends on the absolute value of the various rate constants and thus the efficiency of a muscle will vary with velocity in a complex fashion.

## EFFICIENCY OF ECCENTRIC CONTRACTIONS

Speed and direction of movement are key factors determining the energy costs of muscle contraction. ATP splitting varies with different velocities of shortening and stretching as shown in Figure 2.16. With high velocities of stretch there is very low ATP splitting, because cross-bridges have little opportunity to complete the full cycle and are detached forcibly without requiring ATP binding. Although it may appear that eccentric contractions of this type have a very high efficiency, in reality the concept of efficiency of muscles being stretched is meaningless as no work is generated by the muscle and it simply resists work that is performed upon it.

## EFFICIENCY OF CONCENTRIC CONTRACTIONS
### Experimental protocols

Measuring economy during isometric contractions is, conceptually and practically, relatively easy; it is much more difficult to measure the efficiency of shortening contractions. Even measuring force during shortening is difficult and, when this is combined with attempts to measure energy turnover, the problems are magnified. Energy used in a contraction is most often measured by chemical methods as the difference between a pair of muscles, one stimulated, the other acting as a control. Frozen muscles are extracted and the total ATP turnover calculated from the decrease in ATP and phosphocreatine and the increase in lactate. Most experiments in this area have tended to use an isometric contraction, followed by a release. The problem with this method is that a lot of energy is used during the development of force and the isometric phase, and this does not contribute to the work done during shortening. Consequently, relatively low values for efficiency have been reported (e.g. de Haan et al 1989). With shortening contractions it is important to avoid energy losses due to the stretching and then releasing series elastic elements, and consequently current techniques for measuring efficiency involve muscles shortening under isotonic (or pseudo-isotonic) (e.g. see Fig. 7.10) conditions so that force remains constant during the contraction and there is no change in length of the series elastic components. It is also important to stop the stimulation before the end of shortening to allow the force in the cross-bridges to perform work as the muscle relaxes. Using this experimental approach, considerably higher efficiencies are obtained than previously (de Haan et al 1993).

An alternative way of estimating the energy utilized is to measure heat production during the shortening phase; the mechanical efficiency is the ratio of work output divided by heat plus work. This is an elegant but technically demanding technique which has the advantage that repeated measurements can be made on the same preparation (Curtin & Woledge 1991). The values of mechanical efficiency obtained are similar to those estimated using biochemical methods (Barclay 1994, de Haan et al 1993).

The values for gross efficiency obtained for whole body exercise are around 20%. Considering the anatomical and biomechanical limitations on the transduction of muscle force into external work, it is necessary for muscle efficiency to be about 30% to account for the values observed in whole body exercise. However, the best estimates for the efficiency of isolated muscle preparations are about 21% and it is clear that the difference needs to be explained. There are a number of possibilities, one being that efficiency is greater at lower levels of activation, probably due to a smaller contribution to the total energy costs from the processes of activation (Abbate et al 2002, Buschman et al 1996). Most experiments on isolated preparations use high and constant frequency stimulation, and higher efficiencies might be found if different stimulation protocols were used.

## Effects of shortening velocity and fibre type

Energy turnover, per unit time, is faster during shortening (Fenn effect; see Ch. 2). An individual cross-bridge completes its cycle of attachment, rotation and detachment more rapidly during shortening than during isometric contractions because the relative movement of the actin and myosin filaments carries the cross-bridge into a position where it can detach rapidly in the presence of ATP (see Figs 2.10 & 2.11). If the force that a muscle can generate did not change with velocity, then work done per unit time (power) would increase linearly with velocity. However, we know from the force–velocity relationship that force decreases with increasing velocity so that power first increases, reaches a maximum and then decreases with increasing velocity (see Fig. 2.4). The efficiency of a muscle during shortening is therefore determined by dividing the power curve in Figure 2.4 by the curve for ATP splitting shown in Figure 2.16. The resulting curve is similar in shape to the power curve, with an optimum efficiency at about 25% maximum velocity of unloaded shortening ($V_{max}$) compared with 30% for maximum power (Barclay et al 1993, Heglund & Cavagna 1987). Figure 13.4 shows how efficiency varies with velocity of shortening and the differences between fast and slow fibre types.

The economy of slow fibres has been reported to be 2–4-fold greater than for fast fibres for isometric

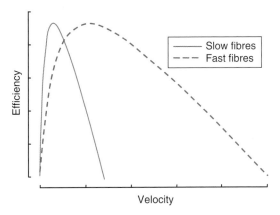

**Figure 13.4** Influence of muscle shortening velocity on efficiency in slow and fast fibres.

contractions. During shortening the *maximum* efficiencies of the two fibre types are similar. However, there is a major difference in the velocity at which the peak efficiency is obtained (Fig. 13.4) and this reflects the fact that $V_{max}$ can be three to four times greater in the fast fibres. In a muscle with a mixed fibre-type composition, the overall efficiency will be determined by the contributions of the different fibres and this will depend critically on the speed of contraction. Thus, in the example given in Figure 13.4, if the contractions are at a low velocity the slow fibres will be working efficiently, whereas at high velocities the same slow fibres will be very inefficient, making virtually no contribution to the force yet still consuming energy.

Data for exercising human muscle is hard to find, although the theoretical implications of differences in muscle fibre-type composition, speed of contraction and effects of temperature have been discussed (Sargeant & Jones 1995).

Competitive cyclists adopt a high pedalling frequency and this has been attributed to a desire to maximize power output and/or efficiency, but there have been relatively few direct measures of efficiency at different pedalling frequencies. Gaesser & Brooks (1975) found efficiency to decrease at higher velocities of cycling, while there was little difference with increases in workload. A little later, Suzuki (1979) showed that the effect of velocity on cycling efficiency depended on the fibre composition of the subjects. Subjects with a predominance of slow-twitch fibres became slightly less efficient when pedalling at the higher cadence, while those described as the 'fast twitch' group became slightly more efficient at the higher velocity. This result is consistent with the relationships shown in Figure 13.4. If the lower pedalling frequency were close to the peak efficiency of the slow fibres, then moving to a higher velocity would decrease the efficiency of the muscles with a predominance of slow fibres. In contrast, moving from a slow to higher velocity would improve the overall efficiency for subjects with mainly fast fibres.

Muscle fibre composition has been shown to affect the efficiency of cycling with efficiencies varying between 19% and 23%, with type 1 fibre content ranging from 40% to 80% (Coyle et al 1992, Horowitz et al 1994). Although apparently small, these differences can have significant effects on performance. In the latter study it was found that subjects with the greater type 1 fibre content had a consistently higher power output over a 1-h performance trial. For exercise at pedal frequencies close to the optimum for the type 1 fibres, the overall efficiency will be a function of the type 1 fibre composition, and for the majority of movements in sporting and everyday life muscle contractions are at relatively low velocities.

Assuming that the peak efficiencies of the slow and fast fibres are the same, it might be thought that carrying out an activity at a low velocity corresponding to the peak of the slow fibre efficiency would be the same as working at a high velocity corresponding to the optimum of the fast fibres. This would be true if it were possible to recruit only slow or only fast fibres at the appropriate velocities. At low velocities and modest workloads, the size principle may dictate that it is predominantly slow motor units that are recruited. In contrast, at high velocities or workloads, both fast and slow units are recruited, and if the movement involves high velocities the slow units will make no contribution to the power yet will still be metabolically active. Consequently the overall efficiency will be reduced, even though the fast units may be working at their optimum. De-recruitment of slow units at high velocities of movement would seem a useful strategy, but there is little evidence that this happens.

## STRETCH–SHORTENING CYCLES

Many common movements involve muscles undergoing cyclical shortening and lengthening, with the lengthening phase being the result of an antagonist muscle contracting or some external force, such as the weight of the body extending the calf and quadriceps during a series of jumps. Attempts have been made to mimic this type of activity with isolated preparations, alternately stretching and releasing the muscles while stimulating at preset intervals. The power output and efficiency obtained in these circumstances varies depending on the precise nature and timing of the stretch and stimulation.

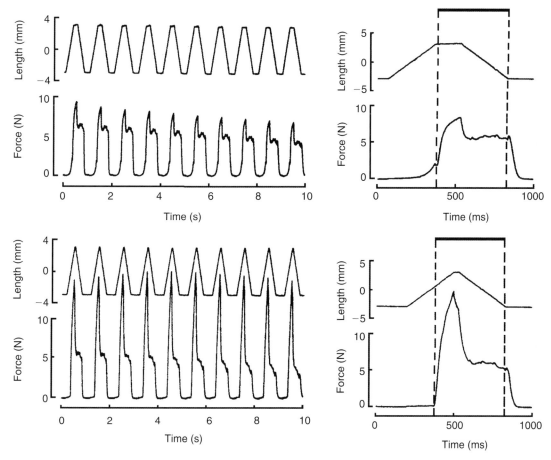

**Figure 13.5** Examples of force and length traces obtained during series of pre-isometric contractions (upper diagrams) and pre-stretch contractions (lower diagrams). Left, series of ten successive contractions. Right, enlarged diagrams of the first contraction of the series. The upper black bars indicate the time of stimulation. Length is given relative to muscle optimum length ($L_o$). (Redrawn from the data of de Haan et al 1989.)

Figure 13.5 shows two different protocols of stretching and release (de Haan et al 1989). In the first, the stretch occurred before the muscle was stimulated and force generation started at the end of the shortening phase, when the muscle was isometric, and continued for 150 ms before shortening began. In the second protocol the muscle was stimulated whilst being stretched and it can be seen that the peak force achieved before the release was much higher. When released, the work done during the shortening phase was about 40% greater than without an active pre-stretch. Most of the additional work was performed shortly after the start of the release and this suggests that the origin of the additional energy liberated was that stored in the series elastic components during the active stretch. High-energy phosphate utilization was very similar in the two types of contraction, so the overall efficiency was 37% with active pre-stretch, compared with 26% without. Similar reports of unlikely high efficiencies are frequently seen in the literature concerned with locomotion using stretch-shortening cycles (Aura & Komi 1986, Heglund & Cavagna 1987). This way of calculating efficiency is potentially misleading, because a significant portion of the work performed by the muscle comes from the external work used to stretch the muscle–tendon complex.

A true energy balance would therefore require subtraction of the work delivered to the muscle from the total work output. When the (net) efficiency is calculated in this way it is always found to be lower than that obtained without an active prestretch (12% versus 24%, in the example shown in Figure 13.5), as it is never possible to recover all the energy put into the muscle during the stretch.

These observations and cautions apply to isolated muscle tendon preparations where external work done on an active muscle should not be represented as a gain in efficiency of the muscle itself. The situation is different when muscles are considered working in the body. In many movements the antagonist muscle is activated towards the end of the movement and acts as a brake to slow down the limb and possibly protect the joint. In the process the antagonist will be stretched whilst active, a situation analogous to that illustrated in Figure 13.5. This energy can be used to increase the subsequent work output of that muscle, provided the movement starts rapidly after the end of stretch. A similar situation applies when a muscle is stretched by gravity, such as when landing from a jump or while running where appreciable quantities of energy are stored in the tendons of the lower leg (van Ingen Schenau et al 1997), and this may explain why the efficiency of running is reported to be greater than that of cycling (Asmussen & Bonde-Petersen 1974). In some animals this is developed to a high degree: horses, dogs, camels and kangaroos all rely on the storage of energy in very long tendons for an efficient running or hopping style (Alexander 1982).

REFERENCES AND FURTHER READING

Abbate F, de Ruiter C J, Offringa C, Sargeant A J, de Haan A 2002 In situ rat fast skeletal muscle is more efficient at submaximal than at maximal activation. Journal of Applied Physiology 92: 2089–2096

Alexander R McN 1982 Locomotion of animals. Glasgow: Blackie

Asmussen E, Bonde-Petersen F 1974 Apparent efficiency and storage of elastic energy in human muscles during exercise. Acta Physiologica Scandinavica 92: 537–545

Aura O, Komi P V 1986 Effects of prestretch intensity on mechanical efficiency of positive work and on the elastic behaviour of skeletal muscle in stretch–shortening cycle exercise. International Journal of Sports Medicine 7: 137–143

Barclay C J 1994 Efficiency of fast- and slow-twitch muscles of the mouse performing cyclic contractions. Journal of Experimental Biology 193: 65–78

Barclay C J, Constable J K, Gibbs C L 1993 Energetics of fast- and slow-twitch muscles of the mouse. Journal of Physiology 472: 61–80

Buschman H P J, Elzinga G, Woledge R C 1996 The effects of the level of activation and shortening velocity on the energy output in type 3 muscle fibres from *Xenopus laevis*. Pflügers Archiv. European Journal of Physiology 433: 153–159

Cavanagh P R, Kram R 1985 The efficiency of human movement—a statement of the problem. Medicine and Science in Sports and Exercise 17: 326–331

Coyle E F, Sidossis L S, Horowitz J F, Beltz J D 1992 Cycling efficiency is related to the percentage of type I muscle fibers. Medicine and Science in Sports and Exercise 24: 782–788

Crow M T, Kushmerick M J 1982 Chemical energetics of slow- and fast-twitch muscles of the mouse. Journal of General Physiology 79: 147–166

Curtin N A, Woledge R C 1991 Efficiency of energy conversion during shortening of muscle fibres from the dogfish *Scyliorhinus canicula*. Journal of Experimental Biology 158: 343–353

De Haan A, Rexwinkel R, van Doorn J E 1988 Influence of muscle dimensions on economy of isometric exercise in rat medial gastrocnemius muscles in situ. European Journal of Applied Physiology 57: 64–69

De Haan A, van Ingen Schenau G J, Ettema G J, Huijing P A, Lodder M A N 1989 Efficiency of rat medial gastrocnemius muscle in contractions with and without an active prestretch. Journal of Experimental Biology 141: 327–341

De Haan A, de Ruiter C J, Lind A, Sargeant A J 1993 Age-related changes in force and efficiency in rat skeletal muscle. Acta Physiologica Scandinavica 147: 347–355

Gaesser G A, Brooks G A 1975 Muscular efficiency during steady-rate exercise: effects of speed and work. Journal of Applied Physiology 38: 1132–1139

Goldspink G 1978 Energy turnover during contraction of different types of muscle. In: Asmussen E, Jorgensen C (eds) Biomechanics VI-A. International Series on Biomechanics, vol 2B, pp 27–39. Baltimore: University Park Press

Heglund N C, Cavagna G A 1987 Mechanical work, oxygen consumption, and efficiency in isolated frog and rat muscle. American Journal of Physiology 253: C22–C29

Homsher E, Kean C J 1978 Skeletal muscle energetics and metabolism. Annual Review of Physiology 40: 93–131

Horowitz J F, Sidossis L S, Coyle E F 1994 High efficiency of type I fibers improves performance. International Journal of Sports Medicine 15: 152–157

Sargeant A J, Jones D A 1995 The significance of motor unit variability in sustaining mechanical output of muscle.

Advances in Experimental Medicine and Biology 384: 323–338

Suzuki Y 1979 Mechanical efficiency of fast- and slow-twitch muscle fibers in man during cycling. Journal of Applied Physiology 47: 263–267

Szentesi P, Zaremba R, van Mechelen W, Stienen G J M 2001 ATP utilization for calcium uptake and force production in different types of human skeletal muscle fibres. Journal of Physiology 531: 393–403

Van Ingen Schenau G J, Bobbert M F, de Haan A 1997 Does elastic energy enhance work and efficiency in the stretch–shortening cycle? Journal of Applied Biomechanics 13: 389–415

## QUESTIONS TO THINK ABOUT

1. If the efficiency of a muscle is given as 20%, what percentage of chemical energy is converted to work output? What happens to the remaining chemical energy?

2. Is it likely that the fraction of energy utilization required for calcium release and uptake differs between fast and slow muscles?

3. Why do isometric contractions of fast muscles use more energy than those of slow muscles?

4. How does (a) the economy and (b) the efficiency of a short muscle compare with that of a long muscle?

5. It is sometimes said that the efficiency of eccentric movements is very high. What do you make of this statement?

6. Draw the velocity–efficiency relations for two muscles with the same volume, one having a cross-sectional area twice that of the other but with fibres half the length.

7. It is reported that efficiency does not change during high-intensity, short-lasting fatiguing exercise. What are the implications of this in terms of calcium release and cross-bridge function?

8. What explanation is there for the fact that the efficiency of cycling is relatively independent of pedal frequency between approximately 60 and 110 r.p.m.?

9. Is the muscle efficiency of stretch–shortening cycles different from that of contractions without an active stretch preceding the shortening?

10. Is high efficiency a cause of high fatigue resistance?

# 14

# Muscle growth, development, adaptations and ageing

In the preceding chapters we have seen how muscle structure, its composition and control are matched to a variety of tasks ranging from the maintenance of posture to the extremes of athletic performance. These functions develop partly under genetic control and partly under the influence of habitual activity and training. Skeletal muscle is a very adaptable tissue in that it responds to the demands placed upon it during growth and training, but it also declines in function with age or immobilization.

## EMBRYONIC ORIGINS AND FETAL DEVELOPMENT

Skeletal muscle is derived from myogenic cells in the embryonic mesoderm. At about the sixth week of gestation these mesodermal stem cells migrate to appropriate sites and begin to differentiate to form myoblasts. Some myoblasts remain as single cells with mitotic potential; these will form the satellite cells of mature muscle. Other myoblasts aggregate and fuse to form multinucleate primary myotubes attached at each end to the developing tendons and skeleton. Within the developing myotubes a central chain of nuclei forms surrounded by basophyllic cytoplasm rich in polyribosomes (Fig. 14.1).

Midway along the primary myotubes further myoblasts aggregate and fuse to form secondary myotubes. At first, the primary and secondary myotubes share a common basement membrane (Fig. 14.2A), but eventually the secondary myotubes develop a separate basement membrane, make contact with the tendon and become independent

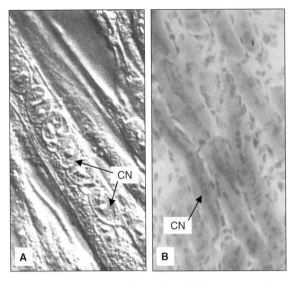

**Figure 14.1** Formation of myotubes. **A**, phase-contrast image of cultured myoblasts fusing to form myotubes with large central nuclei. **B**, fetal muscle showing the chains of dark-staining central nuclei in the newly formed myotubes (haematoxylin and eosin stain). CN, chains of central nuclei.

of the primaries. The proportions of primary and secondary myotubes may vary between muscles and between individuals. Thus it is believed that the soleus, a slow muscle, contains a predominance of fibres derived from primary myotubes, whereas in other faster muscles, such as the extensor digitorum longus, the majority of fibres originate from secondary myotubes. In the human fetus the transition from myoblasts to primary myotube takes place at around the seventh to ninth weeks of gestation, and by the end of this period the primordia of most muscle groups are well defined. At this time the synthesis of the contractile proteins, actin and myosin, begins and the first signs of cross-striation are visible (Fig. 14.2B).

The transition from myoblasts to myotubes is initiated by the expression of a group of muscle transcription factors, products of the *myf* genes (*myf5*, *myoD*, *MRF4* and *myogenin*). This process is inhibited by the presence of *myostatin*, which downregulates *myoD*. Some cattle lack myostatin and this gives rise to muscle hyperplasia and hypertrophy (Langley et al 2002).

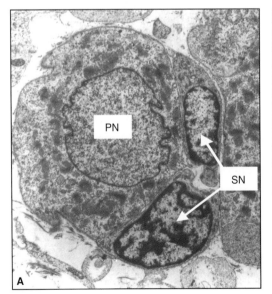

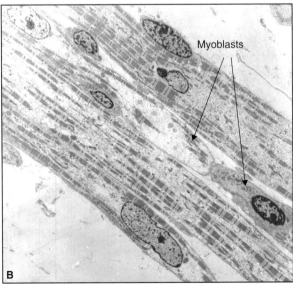

**Figure 14.2** Developing myotubes and fibres. **A**, nucleus of a primary myotube (PN) with two attached secondary myotubes with prominent nuclei (SN). Note that the primary and secondary myotubes are contained within the same basement membrane; 12 weeks' gestation. **B**, longitudinal section of myotubes showing the formation of myofibrils.

From 11 weeks onwards there is a proliferation of myofibrils leading to hypertrophy of the muscle fibres, which also grow in length by the addition of sarcomeres at their ends. At 16–17 weeks a further population of myotubes becomes apparent, known as the tertiary myotubes. These myotubes are small, adhere closely to the secondary myotubes and are enclosed within the same basement membrane. By 18–23 weeks the tertiary myotubes have become independent, while the nuclei of the more mature myotubes begin to move to the periphery of the fibre. Some 5–10% of myoblasts remain as undifferentiated satellite cells (see below).

Developing fibres express a number of different myosins, which can be identified using monoclonal antibodies; these include an embryonic form and an intermediate fast type that confers the properties of the type 2c fibres (often seen in regenerating adult muscle) when stained for myosin ATPase. The primary myotubes can be identified throughout embryonic development as they alone express adult slow myosin from about 9 weeks of gestation (Draeger et al 1987).

At around 10 weeks, the nervous system makes contact with the developing muscle fibres. In response to the contractile activity imposed by the motor nerve, the fibres begin to differentiate so that, eventually, fetal myosins are no longer expressed and about half the fibres express slow myosin and half fast myosin. This process, which is apparent by about 32 weeks of gestation, is probably not fully completed in human muscle until a few months after birth.

## Muscle fibre nuclei

Skeletal muscle is a unique tissue, being composed of fibres that are multinucleated cells. In healthy adult muscle the nuclei lie at the periphery of the fibres just below the plasma membrane (Fig. 14.3A). Viewed with a light microscope, two or three nuclei per fibre can normally be seen in a cross-section of a muscle between 4 and 10 μm thick; thus it can be estimated that there are about 200 to 300 nuclei per millimetre of muscle fibre length.

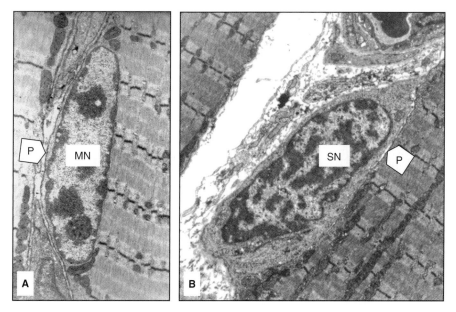

**Figure 14.3** Muscle fibre nucleus and satellite cell. **A**, muscle fibre nucleus (MN) just below the plasma membrane (P). **B**, satellite cell lying outside the muscle plasma membrane with a layer of cytoplasm around the nucleus; SN, satellite cell nucleus.

## Satellite cells

Satellite cells, derived from embryonic myoblasts, lie between the plasma membrane and the basement membrane of muscle fibres. Each satellite cell consists of a large nucleus with a thin layer of cytoplasm. Using an electron microscope, satellite cells can be identified by their position outside the plasma membrane but beneath the basement membrane of the muscle fibre (Fig. 14.3B).

Satellite cells are the source of new nuclear material that is required for muscle growth and hypertrophy and, if the fibre is damaged, satellite cells become activated, divide and fuse to replace the damaged portions of the muscle fibre.

## CONNECTIONS BETWEEN NERVE AND MUSCLE

### First contacts

In the human embryo, probably as early as 10–11 weeks of gestation, axons from motoneurons grow out of the spinal cord and invade the fetal muscle. At first a number of axons form synapses with each embryonic fibre, but as the muscle matures all but one of the synapses are lost (Fig. 14.4). Before innervation the whole membrane surface of the myotube is covered with the fetal form of acetylcholine receptors (AChRs), but with the onset of contractile activity these *extrajunctional* receptors are lost and adult AChRs are restricted to the muscle membrane forming the neuromuscular junction.

The loss of receptors is activity dependent and thought to be a consequence of calcium release in the active fibre switching off AChR gene transcription (Sanes & Lichtman 1999). This raises the question of how AChRs survive under the neuromuscular junction in the mature fibre. Several neuropeptides are released together with acetylcholine from the presynaptic nerve terminal and one of these, *calcitonin gene-related peptide*, is thought to cause the expression of the AChR gene in nuclei immediately under the neuromuscular junction. Thus these nuclei, which can often be seen in the region of the junction (see Fig. 3.1), differ in their gene expression from the majority of nuclei in the fibre.

### Loss of multiple innervation

With nerve muscle preparations maintained in vitro, the rejection process can be prevented by addition of inhibitors such as *leupeptin* that block

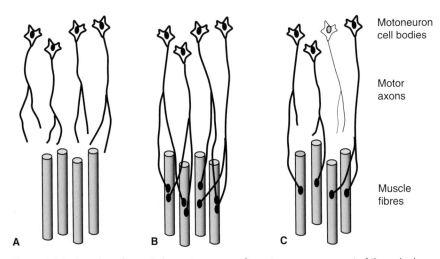

**Figure 14.4** Invasion of muscle by motor nerves. **A**, motor axons grow out of the spinal cord, branch and invade the fetal muscle. **B**, multiple synapses are formed with each fetal muscle fibre. **C**, mature muscle fibres. Each fibre is now innervated by one motor axon branch. One of the motoneurons has failed to make a connection with the muscle fibres and is degenerating.

the action of proteolytic enzymes activated by calcium. The rejection can also be prevented by reducing the calcium concentration in the extracelluar fluid. On the other hand, a high concentration of potassium in the surrounding medium has been shown to speed up the detachment of motor axon branches.

The loss of multiple innervation appears to be brought about by the contractile activity of the developing fibre which, paradoxically, is the result of stimulation by the same attached nerve endings. Contractile activity of the muscle fibre releases potassium ions into the surrounding extracellular space, activating $K^+$-dependent calcium channels in the membranes of nearby axon terminations (Fig. 14.5). Calcium enters the nerve terminal, stimulating calcium-dependent proteases, which attack the structure of the terminal (probably the neurofilaments), causing the axon to retract. The single surviving axonal attachment may be the one that is largest and most well developed; size may be an advantage because it gives a larger ratio of volume to surface area, keeping the calcium concentration relatively low (Vrbová et al 1995). In more mature nerve terminals there might be a change in the number or sensitivity of the $K^+$-dependent calcium channels. Nerve and muscle, therefore, have an uneasy coexistence, the nerve striving to innervate muscle fibres while the muscle fibres appear to do their best to reject those axons that make contact.

The first contacts between nerve and muscle occur between axons and the primary myotubes, leading, at first, to multiple innervation as described above. Later, as the multiple innervation is lost, the secondary myotubes become innervated by the axons that have been rejected by the primary myotubes. This sequence of innervation may have consequences for the later development of different fibre types, as many of the fibres derived from primary myotubes are constrained to become slow type 1 fibres with, apparently, little opportunity for change. The fibres derived from secondary myotubes, however, have the ability to change the expression of a wide range of contractile and other proteins, depending on the pattern of activity imposed on the muscle.

If a muscle fibre becomes denervated, the whole surface membrane again becomes covered with AChRs, and nearby healthy axons sprout, branch and make contact with the fibre. Trophic substances might be released from the denervated fibre to stimulate axonal growth, but the sprouting could also be a permissive process. Active muscle may suppress axonal branching in the same way that it causes the retraction of redundant synapses, and only when the fibre is quiescent might sprouting begin. Several adjacent axons may be induced to sprout and make contact with the denervated fibre, and the process of multiple innervation followed by rejection will occur again, as first took place in the fetal muscle.

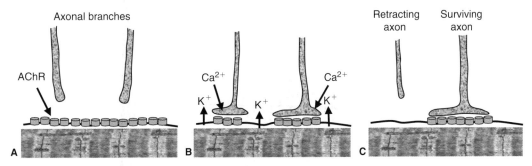

**Figure 14.5** The process of synaptic rejection. **A**, multiple motor axons approach and make contact with the muscle fibre. **B**, activity evoked in the fibre causes $K^+$ release and opening of calcium channels in the presynaptic axonal membrane. **C**, the mature muscle fibre with a single neuromuscular junction and acetylcholine receptors (AChR) restricted to the region of the synapse. The contractile activity in the muscle fibres has two effects: (i) it reduces the number of extrajunctional AChRs and (ii) it causes the rejection of all but one neuromuscular junction.

## MUSCLE GROWTH DURING CHILDHOOD AND ADOLESCENCE

Childhood and adolescence are times of major change in body size, shape and function, and the skeletal muscles are intimately involved in this process.

### Changes in fibre size and character

The number of fibres in each human muscle is probably set by 24 weeks of gestation. In the rat, fibre numbers do not change during life while the mean fibre cross-sectional area increases nearly tenfold from the newborn to adult animal (Rowe & Goldspink 1969). There are considerable practical and ethical problems involved in making measurements of fibre size and number in children, but the limited data available suggest that there is an increase in size without a change in fibre numbers (hypertrophy without hyperplasia) as the muscles grow in size and strength (Fig. 14.6). Adult muscle fibre cross-sectional areas are reached shortly after puberty. In an adult man about 40–45% of the bodyweight is muscle, while the proportion is slightly lower in women. The mean cross-sectional area of fibres in a biopsy from the quadriceps muscle in a normal man ranges from 3500 to $7500\,\mu m^2$ and in normal women from 2000 to $5000\,\mu m^2$ (fibre diameters of about 25–50 $\mu m$).

### Changes in strength

To assess muscle changes during growth it is important to distinguish between improvements in skill and improvements in muscle strength. Performance measures, such as jumping, throwing, etc., involve strength, speed and skill, and in a growing child it is not easy to separate these three. What is required are tests that require only the minimum of skill and, even so, it is difficult to make accurate measurements of strength in children under the age of 5–6 years. However, after this age, reliable measurements can be made and there is a steady and similar increase in strength in both sexes up to the age of puberty when, during the growth spurt and sexual maturation, a more rapid increase occurs (Round et al 1999) (Fig. 14.7).

In the years before puberty, muscles of the lower limbs grow in proportion to the bodyweight (or height[3]). Teleologically this might be

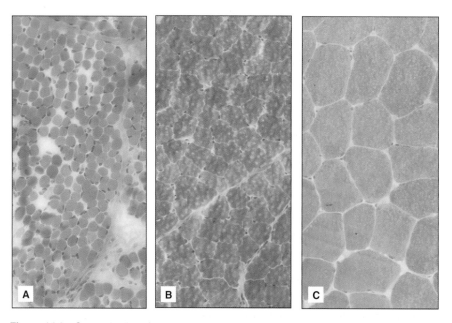

**Figure 14.6** Growth in size of human muscle fibres. **A**, baby aged 8 months. **B**, boy aged 5 years. **C**, boy aged 14 years. Quadriceps muscle, all at the same magnification.

expected as leg muscles have the function of bearing the bodyweight. Muscles in the upper body, however, grow less rapidly and do so in proportion to height[2], the muscles maintaining the same proportions of length to breadth as the long bones increase in length. However, during puberty and the adolescent growth spurt, the skeletal muscles grow more rapidly in relation to height and bodyweight.

For adolescent girls and adult women, the proportions of muscle strength to bodyweight and height remain much the same as those for younger children. In adolescent boys, however, there is an additional stimulus for muscle growth that is particularly noticeable in the muscles of the upper limb girdle and is associated with the large increase in circulating testosterone levels that marks the transition from boyhood to the awkward maturity of the late teenage years (Round et al 1999).

Most men can manage at least one press-up or chin-up, whereas these exercises are very difficult for the majority of women. Throwing, pole vaulting and gymnastic exercises involving strength of the upper body are other activities where men have a clear advantage over women. It is not clear, however, what evolutionary advantage is gained by one sex having particularly strong arms. The most likely explanation is that it is of little practical advantage but that the upper body muscle development is another male secondary sexual characteristic. The heavy musculature may originally have served the function of either attracting females or of impressing and dominating other m... ter being the more likely explanation cons... the behaviour of most primates and other ma... malian species such as cattle and deer.

## AGEING AND MUSCLE FUNCTION

Starting as early as 30 years but becoming significant around the age of 40 years, there is a progressive loss of muscle mass and decrease in strength. The loss of muscle function in the seventh decade onwards is associated with increased susceptibility to falling, a general decrease in mobility and frequently a loss of independence. With a rapidly ageing population this presents a serious problem for the health and social services, and an understanding the underlying cause of the muscle changes is the first step to finding ways of minimizing the problem. In women the loss of force is particularly marked in the years after the menopause when there are very similar decreases in the mass of both muscle and bone as a consequence of the loss of oestrogen (Rutherford & Jones 1992). The loss of muscle in elderly subjects is progressive, and between the ages of about 50 and 90 years muscle mass may be reduced by 50% (Fig. 14.8A). There are two aspects of the muscle changes. The first is a decrease in isometric strength which, in part, is due to general muscle fibre atrophy, although the total muscle cross-sectional area does not decrease as much, probably because of an accumulation of fat and connective

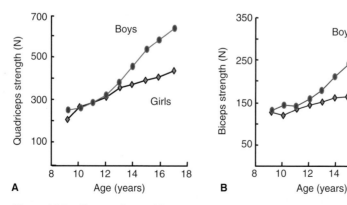

**Figure 14.7** Changes in quadriceps and biceps muscle strength with age. Note the steady increase before puberty and, for boys, a rapid increase during and for a while after puberty. (Redrawn from the data of Round et al 1999.)

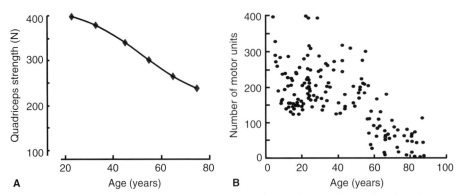

**Figure 14.8**  Muscle changes with age. **A**, quadriceps isometric strength as a function of age in healthy women. (Data from Rutherford & Jones 1992.) **B**, loss of functioning motor units of the extensor digitorum brevis muscle with age in healthy individuals. (Redrawn from the data of McComas 1996.)

tissue (Rutherford & Jones 1992). The second problem is slowing of the contractile properties of the muscle and a loss of power which is greater than can be explained by the loss of isometric force. The implication is that there is a loss of fast myosin which may, in part, be due to a preferential atrophy of type 2 fibres. However, the main cause is motoneuron cell death in the anterior horn of the spinal cord; it is the larger motoneurons supplying the fast motor units that are most affected. Neuronal cell death throughout the central nervous system is a feature of ageing and it is not surprising that motoneurons are also affected.

Loss of motoneurons means a loss of motor units, and this can be demonstrated in small muscles of the hand and foot, where it is possible to estimate the numbers of motor units by counting the force increments while gradually increasing the stimulating current (Fig. 14.8B). What is interesting, and unexplained, is the fact the larger motoneurons innervating fast motor units should be particularly susceptible. The consequences of the loss of innervation to the fast motor units can be seen in biopsies of elderly muscle as areas of fibre-type grouping (Fig. 14.9). This comes about as the surviving axons, which tend to support slow units, sprout and reinnervate the formerly fast fibres. These fibres then take on the contractile and biochemical characteristics of the other fibres innervated by the slow motoneuron.

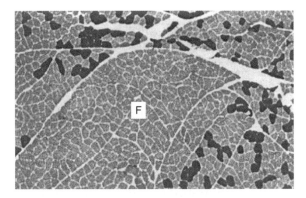

**Figure 14.9**  Adductor pollicis muscle of the hand (autopsy sample, transverse section) from an elderly woman with no clinical neurological signs. F, fascicle containing only type 1 fibres indicating reinnervation (ATPase pH 9.4, type 2 fibres stain dark).

There is no evidence that habitual exercise can help to prevent the major changes in muscle, which are a consequence of motoneuron cell death. However, elderly muscle will respond to strength training, much as younger muscle does, so some of the changes in strength can be offset, although this will not affect the speed of the muscle and has only limited benefits in terms of power. Regular exercise will also help maintain flexibility and, possibly, balance. Moreover, regular exercise is the major way of minimizing the risks of obesity, diabetes and cardiovascular disease that are the major cause of death in the Western world.

# INFLUENCE OF CHRONIC ACTIVITY

Muscle fibres and motor units fall into groups distinguished by their biochemical and physiological properties. As it is these properties that determine the use that can be made of a muscle, it is of great interest to understand how gene expression, which determines the muscle fibre type, is controlled.

Top endurance athletes typically have more than 80% type 1 fibres in their leg muscles and there is considerable debate as to whether they were born that way or achieve the fibre-type disproportion as a result of prolonged training. Studies of identical twins have shown that there is a strong genetic influence over fibre-type proportions, as there is for maximum aerobic exercise capacity ($VO_{2max}$), which itself is determined, in part, by the fibre-type composition.

The first clue as to the mechanism of muscle plasticity came almost by accident in a series of experiments by Buller et al (1960), who were interested in the adaptability of the central nervous system. Motor nerves supplying muscles at the front and back of a cat's leg were transplanted so that the nerve normally supplying the soleus at the back was inserted into the extensor digitorum longus (EDL) at the front, and vice versa (Fig. 14.10A). Time was allowed for the nerves to regenerate and innervate the muscles, and the intention was to see how well the central nervous system could cope with the situation, learning to activate the muscles in a new sequence. The results were disappointing in as much as the central nervous system was unable to deal with such disruption, but there proved to be an unexpected and very important new finding. The soleus at the back of the leg is normally a slow red muscle, while the extensor digitorum longus at the front is a white fast muscle. However, at the end of the experiment it was noticed that the cross-reinnervated muscles had changed colour: the soleus was now pale and the EDL was red. When electrically stimulated it was found that the muscles had also changed their contractile characteristics from slow to fast, in the case of the soleus, and from fast to slow for the EDL (Fig. 14.10B).

From the experiment described above it is clear that the nerve supplying the muscle determines

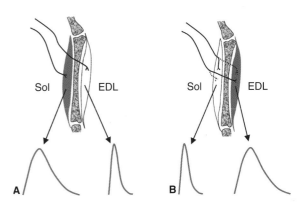

**Figure 14.10** Effect of cross-innervation on contractile properties. **A**, before operation with the slow soleus and fast extensor digitorum longus (EDL). **B**, after cross-innervation. Lower traces show the twitch response of the muscles.

the histochemical and contractile properties of muscle fibres. The question then arose as to what feature of the innervation was important. Was some growth factor passing down the nerve from the motoneuron, or was the determining factor the activity pattern that the nerve imposed on the muscle?

Salmons & Vrbová (1969) implanted electrodes around the motor nerve serving a fast muscle (EDL) in the rabbit and stimulated at low frequency for several hours a day. After 30 days this fast muscle came to resemble the slow soleus muscle in its appearance and contractile characteristics. The same workers later demonstrated that the presence of the nerve itself is not required as, if conduction along the nerve was blocked and the muscle stimulated directly, the change from fast to slow characteristics still occurred. These experiments indicate that it is the pattern of activity imposed on the muscle, rather than trophic substances coming from the nerve, that regulates the expression of genes coding for proteins responsible for the slow contractile properties.

The rationale for using prolonged low-frequency stimulation was that this reflects the type of electrical activity that can be recorded from slow muscles. In general, slow muscles are used for minor adjustments of posture and position, and are consequently required to be continuously active, making frequent small contractions while the animal is awake. In contrast, fast muscles have a pattern of activity whereby they are electrically silent for long periods and then discharge with short high-frequency bursts as the animal makes occasional rapid or high-force contractions. However, it transpires that it is not the frequency itself that is of prime importance but rather the total duration of the activity, so that stimulation protocols that include a high-frequency component are equally effective. Indeed, including occasional bursts of high-frequency stimulation helps to maintain muscle size and strength, which otherwise tends to decrease with endurance training.

The major changes seen after chronic stimulation include an increase in capillary density, proliferation of mitochondria, decrease in sarcoplasmic reticulum and $Ca^{2+}$-ATPase, and the expression of different myosin and troponin isoforms. These changes all occur with different time courses (Fig. 14.11). Increased capillary density and mitochondrial content, which are amongst the first changes to be seen in response to prolonged activity, will predominantly affect the fatiguability of the muscle. Changes in the sarcoplasmic reticulum and contractile proteins, which require more activity, influence the speed of the muscle.

If low-frequency stimulation is applied to rabbit muscle for 24 h a day, the total transformation process is complete in about 8 weeks. At that point, the originally fast muscle fibres have been transformed into slow fibres, which are virtually indistinguishable from normal slow fibres; this includes features such as thickness of Z lines and volume of the T tubules. However, rabbit muscle seems particularly easy to transform, whereas rat and mouse muscle take longer and, possibly, may never transform fully. Pette (1998) noted that after 60 days' chronic stimulation of rat EDL there was a major shift in myosin isoform expression from 2b and 2x towards 2a, but there were only 10% type 1 slow myosin heavy chains. Thyroid hormones may play an important modulating role because it is easier to transform fast into slow muscle in hypothyroid states.

The speed of transformed fast muscle decreases, as indicated by slower rates of force development and relaxation, and a shift of the force–frequency

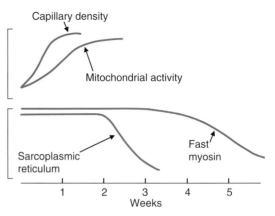

**Figure 14.11** Time course of changes with prolonged activity in fast rat or rabbit muscle. Changes in capillary density, mitochondrial content, sarcoplasmic reticulum and fast myosin ATPase as a result of chronic stimulation.

relationship to the left and lower frequencies of stimulation. The fatigue resistance of the muscle becomes very similar to that of a highly oxidative muscle.

Although prolonged activity imposed by the nervous system has a major influence on the expression of fibre characteristics, not all fibres are susceptible to change (Unquez et al 1993); these may be the fibres derived from the primary myotubes. In the past it has been thought that imposed activity is the controlling influence for the expression of all the genes involved in setting the speed of a muscle. A more considered judgement is that activity can modulate genetic expression within certain limits so that, for instance, for the myosin subtypes, 2x fibres express increased amounts of 2a with prolonged activity but not type 1 myosin, while type 2a fibres begin to express type 1 myosin.

## Human endurance training

Although there has been a great deal of work on stimulated animal muscle, there are relatively few investigations on the effects of chronic stimulation on normal human muscle. Studies that have been undertaken show that changes in fatiguability can be produced but that the stimulation has probably never been sufficient to alter the myosin isoform expression, converting type 2 fibres to type 1 (Rutherford & Jones 1988). Stimulation for longer periods of up to 6 months in patients with spinal cord injuries has shown some evidence of a transformation of fast to slow myosins as well as improvements in oxidative enzymes (Martin et al 1992). Stimulation for 12 months resulted in an increase in type 2a myosin, from 20% to over 90% (Andersen et al 1996).

One of the difficulties with voluntary exercise is that during endurance training only part of the muscle is used. Fibre recruitment depends on the intensity and duration of the training; the intensity determines which motor units are recruited and, as duration increases, the first motor units to be recruited eventually fatigue and faster units have to make up the deficit. In contrast, electrical stimulation activates all the fibres irrespective of fibre type and this may account for the fact that fibre-type transformations can occur in 6–12 months with electrical stimulation, whereas it may take many years of athletic training to accomplish this. Nevertheless, the adaptive changes of muscles during physical training are qualitatively similar to those during chronic stimulation. The adaptations gained during training are (unfortunately) easily and rapidly reversible. Increases in oxidative enzyme activities during endurance training may take months, while there is a fast decline of these activities within a few weeks after stopping the training.

## Mechanisms of adaptation to increased activity

Chronic activity, whether voluntary or imposed by electrical stimulation, leads to many changes in muscle, ranging from increased capillary density to changes in myosin expression. It is tempting to think of a set of genes turned on by a single switch, but there is no evidence that this is the case and there are many reasons to believe that fibre-type expression is determined by multiple genes each responding to different stimuli arising from activity. Although there are many candidates for the link between activity and gene expression, none has so far been clearly identified. Capillary growth in response to increased activity is stimulated, in part, by the mechanical action of increased blood flow through existing capillaries. Increased pressure on capillary walls stimulates budding and branching of the capillaries, leading to an increased capillary bed. Mechanical activity in the muscle and growth factors such as *vascular endothelial growth factor* (VEGF) are also involved (Hudlicka 1998).

Training and hypoxia are potent stimuli for mitochondrial growth and increase in muscle oxidative capacity. Very little is known about the process, but hypoxia induces *hypoxia-inducible factor 1* (HIF-1), VEGF and heat shock proteins, all of which could play a role in stimulating mitochondrial proliferation (Vogt et al 2001). There remains the question of what aspect of activity causes the production of these factors; decreased glucose and raised lactic acid levels have both been proposed and there is some evidence that the products of glycolysis play a role. Patients with mitochondrial disorders experience high levels of muscle acidosis and have a proliferation of mitochondria in their muscles. In contrast, patients with McArdle's disease have major metabolic problems but do not accumulate lactate and do not have an abnormal mitochondrial content in their muscles.

Prolonged increases in activity will lead to chronic changes in the level of muscle metabolites, altering the ratios of ATP to ADP and $P_i$ and these changes occur as a result of both hypoxia and exercise. However, whilst hypoxia is a potent stimulus for increased mitochondrial content, it is not associated with change in myosin expression (Green 1992). Prolonged activity leads to a rise in the average concentration of calcium within the fibre and there is evidence that free calcium influences the expression of calcium transport ATPase in the sarcoplasmic reticulum (Muller et al 1992). However, there is no evidence that calcium is involved in regulating the expression of myosin isoforms (Pette & Düsterhöft 1992).

In addition, it is important to bear in mind that other factors, such as the hormonal environment (especially thyroxine) and muscle length, also have an influence on gene expression.

One important adaptation to endurance training or chronic stimulation is that the muscle fibres become smaller in cross-section. At first sight this appears somewhat anomalous in that more activity leads to weaker muscles, but the advantage of smaller fibres is that the diffusion distance for oxygen is reduced and the muscle is able to sustain oxidative metabolism for longer. Cardiac muscle, which is the ultimate oxidative endurance muscle, has fibres that are very much smaller than those in the average skeletal muscle.

## OVERLOAD AND STRENGTH TRAINING

Experimental animal models of muscle strength training are difficult to set up because, although it is possible to get rats to run on a treadmill to simulate endurance training, it is much more difficult to persuade them to lift weights. However, one model that has been extensively used is that of compensatory hypertrophy induced by ablation of synergistic muscles (Baldwin et al 1982). If the gastrocnemius and soleus muscles of a rat triceps surae are removed, the remaining plantaris muscle will be overloaded, having to cope with the loads normally shared with the other muscles. The plantaris is consequently exposed to a higher load than normal, and this leads to hypertrophy of the muscle fibres and increase of force. However, not only is the muscle required to generate greater forces, it is also used more frequently, and this provides a higher metabolic load on the muscle, similar to that experienced as a result of chronic stimulation. It is therefore not surprising that the muscle also changes its composition to slower and more oxidative fibres.

Strength training can occur as a result of heavy physical labour or deliberate training for sporting events, or as an end in itself for those interested in body building. The area is one of considerable interest and controversy, and a full treatment is beyond the scope of this chapter. The interested reader is referred to recent reviews (e.g. Jones & Folland 2001) and only a brief discussion of the major points will be given here.

Individuals who work with weights, or who use more modern training machines, become stronger and are able to lift larger weights. In addition, over a period of time their muscles

become larger, which is frequently the individual's reason for weight training in the first place. The main points of interest and controversy can be summarized as follows:

1. Whilst muscle cross-sectional area is the major determinant of strength, the gains in strength seen as a result of training are not necessarily matched by an increase in muscle size, especially in the early stages.

2. It is not known what form of training produces the best results. It is not clear, for instance, whether it is a good idea to fatigue the muscle during training, or whether damaging the muscle leads to greater compensatory growth. The stimulus for muscle growth as a result of weight training is not known.

3. Muscle growth occurs by hypertrophy of individual fibres rather than the appearance of new fibres. This can occur in two ways, either by stimulating the activity of existing nuclei within the muscle fibre, or by causing satellite cells to divide and become incorporated into the fibres. The relative contributions of these two pathways are not known.

Published work on strength training ranges widely from the websites of enthusiasts and publicity material from manufacturers to journal articles concerned with the details of gene regulation. In all of this it is sometimes difficult to make comparisons, and basic terms such as 'strength' are often used in different ways. When considering strength, a distinction needs to be made between the generation of a maximum isometric contraction at optimum length, lasting 2–3 s, which is probably the best measure of the total contractile material in a muscle (i.e. the number of cross-bridges arranged in parallel), and other measures such as isokinetic strength or the ability to lift weights.

The factors that affect isometric strength, such as the angle of pennation, frequency of activation, etc., have been discussed elsewhere (see Ch. 7) The factors that influence isometric force also affect isokinetic contractions but, in addition, isokinetic performance is in part determined by the speed of the muscle and its length. It is also necessary to activate the muscle rapidly, as most dynamic movements last for only a fraction of a second. Knee extension at 90° per second takes 1 s, at most, and this is normally regarded as a relatively slow movement. The maximum training weight that can be lifted (normally expressed as 1 repetition maximum, 1 RM) is another commonly used measure of strength, and this involves all the factors associated with isokinetic measures together with a number of other variables such as the level of skill and coordination. Lifting free weights involves the coordination of many muscles throughout the body, both as synergists for the main load-bearing movement and to stabilize the trunk and shoulder.

Discrepancies between gains in isometric strength and muscle size are probably explained by changes in fibre pennation, decreases in intramuscular fat and increased connective tissue attachments (Jones et al 1989), and by the fact that most measurements of muscle size are made on relaxed muscle whereas muscles change their shape and relationship to the joints over which they act, when active. For major muscle groups, it is unlikely that there are changes in the ability to activate the muscles during isometric contractions with training, as it can be demonstrated by electrical stimulation that voluntary activation is high to start with and does not change. With dynamic contractions, and especially those involving skilful movements, however, it is very likely that changes in the pattern of neural activation play a significant part in increasing performance. There is evidence that a period of training with fast shortening contractions leads to changes in the pattern of neural activation so the muscle is stimulated with a series of doublets at the start of the contraction, giving a rapid rate of force development (Van Cutsem et al 1998).

There is a common belief that it is important to push the muscle to the point of fatigue when weight training, and this would make some sense if it were clear that a build-up of metabolites was the critical factor stimulating muscle growth. This is not the case, however, as very little difference is seen between training protocols that fatigue the muscle with a series of high-force contractions and those where the same amount of work is undertaken but sufficient time is left between contractions for metabolites to recover

fully. Likewise, some body builders believe that damaging the muscle leads to compensatory hypertrophy, but, again, there is little evidence to substantiate this view. It is probably the oedema caused by the damage that gives the impression of a larger, well defined, muscle—which is the main objective for most body builders.

The only consistent message from the research literature is that gains in muscle strength and size occur in response to some combination of high forces and total duration of activity. Thus, the greater the weights and the more frequent the training, the greater the benefits. Heavy weights are probably important both as a way of providing an appropriate mechanical stimulus and also because they ensure that all motor units are recruited and therefore subjected to the training stimulus. Small weights and high repetitions provides a formula for improving the endurance characteristics of the small, slow, motor units.

There is increasing interest in intracellular signalling pathways that are sensitive to mechanical events in muscle as a way of stimulating growth and in particular stretch-sensitive calcium channels. A complex cytoskeletal system, which includes dystrophin (see Fig. 1.5), links the contractile apparatus to the surface and basement membrane. This is associated with nitric oxide synthase generating nitric oxide, which is a potent intracellular messenger. It is possible that this, or a similar system, may stimulate protein synthesis within the fibre and/or lead to the production of growth factors such as insulin-like growth factor 1, which could stimulate satellite cell division and thereby increase the amount of nuclear material within the fibre.

## ADAPTATIONS TO DECREASED MUSCLE USE

There are several models that can be used to study the effects of decreased use on skeletal muscle function.

### Immobilization

Immobilization, such as by plaster casting a leg or arm, results in rapid changes in muscle size and function, as anyone who has had a broken bone will remember. It might be argued that the loss of muscle is a consequence of damage to bone itself, but healthy non-injured subjects immobilized with a plaster around the knee show up to 40% loss of thigh volume within 6 weeks. The time course of this loss is not clear, but it is likely that most occurs within the first week or two. By 6 weeks (when the plaster usually comes off for the first time) muscle biopsy shows that the atrophy is largely explained by shrinkage of individual muscle fibres. In animal models the atrophy is similar in fast and slow fibres, but in human muscle there is a preferential loss of the slow type 1 fibres. The loss of slow fibre area can be dramatic, there being reports of a decrease from 81% of the muscle area to 58% in 6 weeks in one highly trained competitive cross-country skier. In another athlete the injured leg contained 20% type 1, by area, while the other leg contained 69% (Haggmark et al 1986). Although atrophy of the type 1 fibres occurs, there is little evidence to suggest any transformation of slow to fast myosin heavy chains during this timeframe. Another common model of immobilization is bedrest, where similar changes are seen after 6 weeks, although they are not so marked as with immobilization in plaster (e.g. Berg et al 1997).

The length at which the muscle is held is important. Immobilization in a shortened position tends to accentuate the losses, whereas if the muscle is held in a lengthened position mass and function are better preserved.

Immobilization with a plaster cast does not totally inhibit muscular activity. For biarticular muscles such as the rectus femoris, which passes over the knee and hip, the muscle may be quite active and its size and function are preserved because of movements around the hip even though the knee is immobilized.

Another model is hypogravity or unloading, in which the hind limbs of rats or mice are suspended so that they are no longer weight-bearing. The extent of atrophy of individual muscles seen in these situations depends on the extent to which the muscle was used for weight-bearing in normal circumstances: the more active the muscle, the more susceptible it is to atrophy. For example, Thomason et al (1987) found that after 3 weeks of

unloading, plantaris muscle weight decreased by only 17%, whilst the soleus fell by 60% of its normal weight.

## Cordotomy

Cordotomy, both in animal experiments and as a result of traumatic injury to the spine in human patients, is a more drastic form of immobilization and has proportionately greater effects on muscle size and function compared with the models discussed above. In addition to muscle atrophy there are marked changes in rates of contraction and relaxation and in fibre-type composition.

Although there is a good deal of variation, quadriceps force of the spinal cord-injured (SCI) patients is reduced to about 50%. When SCI muscle is stimulated at a low frequency the oscillation of force relative to the mean force is greater than that seen with intact control subjects (Gerrits et al 1999) (Fig. 14.12). There is also a faster rate of force development and relaxation, which all suggest that the SCI muscle is faster. This is confirmed by examining the fibre-type composition of muscle biopsies, where it is seen that there has been a major change towards faster fibres (see Ch. 15, Fig. 15.2; Round et al 1993).

As would be expected, transformed SCI muscle is very susceptible to fatigue during electrical stimulation. As SCI subjects have no voluntary

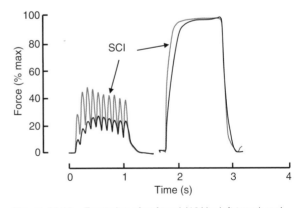

**Figure 14.12** Examples of unfused (10 Hz; left trace) and fused (100 Hz; right trace) contractions of the quadriceps of a subject with a spinal cord injury (SCI) and a control subject (lower trace in each record). Force has been normalized for the maximal force at 100-Hz stimulation. (Redrawn from the data of Gerrits et al 1999.)

control of their muscles below the lesion, the rapid fatigue might not be thought to be a problem. However, there is increasing interest in the use of functional electrical stimulation to help SCI subjects stand and, in some cases, walk. Stimulated cycling is a valuable way to improve cardiovascular function and control of blood pressure, but a limiting factor is often the rapid onset of fatigue. An important line of research is devoted to finding stimulation protocols that will minimize fatigue together with training regimens that will increase fatigue resistance while building up muscle bulk.

The muscle changes seen in the disuse models discussed above indicate that if muscles are not used the contractile and metabolic properties change towards faster muscle types. The implication of this observation is that the fast muscle fibre type is the default setting of the genetic regulatory system. Not only is this of importance in understanding the normal biochemistry and physiology of muscle, but it also has important implications for training, where it becomes difficult to know what to advise to improve sprint performance. It appears that the most effective way of increasing the expression of fast myosin is to immobilize the muscle, but this leads to a loss of muscle mass, which has obvious disadvantages. The only effective way of increasing power, therefore, is to increase muscle bulk, and the increases in muscle force that result from this outweigh any decrease in muscle speed (Jones et al 1989).

## REFERENCES AND FURTHER READING

Andersen J L, Mohr T, Biering-Sorensen F, Galbo H, Kjaer M 1996 Myosin heavy chain isoform transformation in single fibres from m. vastus lateralis in spinal cord injured individuals: effects of long-term functional stimulation (FES). Pflügers Archiv. European Journal of Physiology 431: 513–518

Baldwin K M, Valdez V, Herrick R E, MacIntosh A M, Roy R R 1982 Biochemical properties of overloaded fast-twitch skeletal muscle. Journal of Applied Physiology 52: 467–472

Berg H E, Larsson L, Tesch P A 1997 Lower limb skeletal muscle function after 6 wk of bed rest. Journal of Applied Physiology 82: 182–188

Buller A J, Eccles J C, Eccles R M 1960 Interactions between motor neurons and muscles in respect of the characteristic speeds of their responses. Journal of Physiology 150: 417–439

Draeger A, Weeds A G, Fitzsimmons R B 1987 Primary, secondary and tertiary myotubes in developing muscle: a new approach to the analysis of human myogenesis. Journal of the Neurological Sciences 81: 19–43

Gerrits H L, de Haan A, Hopman M T E, van der Woude L H V, Jones D A, Sargeant A J 1999 Contractile properties of quadriceps muscle in individuals with spinal cord injury. Muscle and Nerve 22: 1249–1256

Green H J 1992 Muscular adaptations at extreme altitude: metabolic implications during exercise. International Journal of Sports Medicine 13 (Suppl 1): S163–S165

Haggmark T, Eriksson E, Jansson E 1986 Muscle fiber type changes in human skeletal muscle after injuries and immobilization. Orthopedics 9: 181–185

Hudlicka O 1998 Is physiological angiogenesis in skeletal muscle regulated by organ microcirculation? Microcirculation 5: 5–23

Jones D A, Folland J P 2001 Strength training in young adults. In: Maffuli N, Chan K M, Macdonald R, Malina R M, Parker A W (eds) Sports medicine for specific ages and abilities. pp 57–64. Edinburgh: Churchill Livingstone

Jones D A, Rutherford O M, Parker D F 1989 Physiological changes in skeletal muscle as a result of strength training. Quarterly Journal of Experimental Physiology 74: 233–256

Langley B, Thomas M, Bishop A, Sharma M, Stewart G, Kambadur R 2002 Myostatin inhibits myoblast differentiation by down-regulating MyoD expression. Journal of Biological Chemistry 277: 49831–49840

Martin T P, Stein R B, Hoeppner P H, Reid D C 1992 Influence of electrical stimulation on the morphological and metabolic properties of paralyzed muscle. Journal of Applied Physiology 72: 1401–1406

McComas A J 1996 Skeletal muscle; form and function. Champaign, Illinois: Human Kinetics

Muller A, van Hardeveld C, Simonides W S, van Rijn J 1992 $Ca^{2+}$ homeostasis and fast type sarcoplasmic reticulum $Ca^{2+}$-ATPase expression in L6 muscle cells. Role of thyroid hormone. Biochemical Journal 283: 713–718

Pette D 1998 Training effects on the contractile apparatus. Acta Physiologica Scandinavica 162: 367–376

Pette D, Düsterhöft S 1992 Altered gene expression in fast-twitch muscle induced by chronic low frequency stimulation. American Journal of Physiology 262: R333–R338

Round J M, Barr F M D, Moffat B, Jones D A 1993 Fibre areas and histochemical fibre types in the quadriceps muscle of paraplegic subjects. Journal of Neurological Sciences 116: 207–211

Round J M, Jones D A, Honour J W, Nevill A M 1999 Hormonal factors in the development of differences in strength between boys and girls during adolescence: a longitudinal study. Annals of Human Biology 26: 49–62

Rowe R W D, Goldspink G 1969 Muscle fibre growth in five different muscles in both sexes of mice: I. normal mice. Journal of Anatomy 104: 519–530

Rutherford O M, Jones D A 1988 Contractile properties and fatiguability of the human adductor pollicis and first dorsal interosseus: a comparison of the effects of two chronic stimulation patterns. Journal of Neurological Sciences 85: 319–331

Rutherford O M, Jones D A 1992 The relationship of muscle and bone loss and activity levels with age in women. Age and Ageing 21: 286–293

Salmons S, Vrbová G 1969 The influence of activity on some contractile characteristics of mammalian fast and slow muscles. Journal of Physiology 201: 535–549

Sanes J R, Lichtman J W 1999 Development of the vertebrate neuromuscular junction. Annual Review of Neuroscience 22: 389–442

Thomason D B, Herrick R E, Surdyka D, Baldwin K M 1987 Time course of soleus muscle myosin expression during hind limb suspension and recovery. Journal of Applied Physiology 63: 130–137

Unguez G A, Bodine-Fowler S C, Roy R R, Pierotti D J, Edgerton V R 1993 Evidence of incomplete neural control of motor unit properties in cat tibialis anterior after self-reinnervation. Journal of Physiology 472: 103–125

Van Cutsem M, Duchateau J, Hainaut K 1998 Changes in single motor unit behaviour contribute to the increase in contraction speed after dynamic training in humans. Journal of Physiology 513: 295–305

Vandervoort A A, McComas A J 1986 Contractile changes in opposing muscles of the human ankle joint with ageing. Journal of Applied Physiology 61: 361–367

Vogt M, Puntschart A, Geiser J, Zuleger C, Billeter R, Hoppler H 2001 Molecular adaptations in human skeletal muscle to endurance training under simulated hypoxic conditions. Journal of Applied Physiology 91: 173–182

Vrbová G, Gordon T, Jones R 1995 Nerve–muscle interactions. London: Chapman & Hall

## QUESTIONS TO THINK ABOUT

1. During childhood and adolescence there are large increases in muscle size and strength. In what way does the growth of muscle differ from the growth of other organs, such as the liver?

2. In running events the difference in performance between men and women is usually of the order of 10%, whereas in the javelin and shot men can throw twice as far as women. Why is this?

3. Fast fibres are used only occasionally to generate high forces and power, and might therefore be regarded as highly specialized fibres with the slow fibres being the normal expression. Is this the correct view of the control of fibre-type expression?

4. How was it shown that the influence of nerve on muscle contractile properties was not due to trophic substances released from the nerve?

5. Is there likely to be a single genetic switch that changes muscle from fast to slow as the result of chronic stimulation?

6. What aspects of muscular activity might initiate the transition from fast to slow characteristics and is there any evidence to support any particular cause?

7. Explain why isomeric force is determined by the cross-sectional area of a muscle. Is this also true during strength training, and what explanations have been suggested for any departure from this relationship?

8. What is the effect of immobilization as a result of a limb in plaster on different fibre types in skeletal muscle?

9. How do the results of immobilization compare with changes seen in patients with spinal cord injury?

10. Is the loss of muscle with ageing simply the result of disuse? If not, what are the reasons?

# 15

# Muscle diseases

Muscle diseases present in many different ways. For a child there may be a delay in achieving the normal motor milestones such as sitting and walking; an adult may not be able to rise from a chair, climb stairs or turn over in bed, or may experience excessive fatigue. Patients can generally be divided into three groups: (1) those who are weak even when rested; (2) those who are of normal strength at rest but for whom even a small amount of exercise leads to premature or excessive fatigue; and (3) those who are not necessarily weak or especially easily fatigued but have some disturbance of function such as spasticity, inability to relax the muscle after contraction (myotonia), abnormally severe cramps or episodes of weakness. Finally, there is a group of patients who complain of excessive fatigue, often associated with ill-defined muscle aches and pains, but in whom little evidence of altered muscle function can be found.

This chapter can give only an outline of the major disease categories affecting skeletal muscle and the interested reader is referred to larger specialist works for more detail; these are listed under References and further reading.

The primary cause of muscle symptoms may be a defect in the muscle itself (myopathy) or may be secondary to changes in the nervous system innervating the muscle (neuropathy).

## MYOPATHIES

Myopathies may be classified in a number of ways, and one convenient way is to consider first whether or not the problem arises as a result of the loss of muscle mass.

## Loss of muscle mass

Muscle bulk may be reduced either as a consequence of a reduction in the size of individual fibres, with the numbers of fibres remaining relatively constant (atrophic myopathy), or fibres may be damaged and destroyed so that the number of fibres in a muscle is reduced (destructive myopathy) (Fig. 15.1).

### Atrophic myopathies

A reduction in muscle fibre cross-sectional area, or atrophy, is frequently a secondary feature of an underlying disease or metabolic stress on the body, and these muscle disorders are sometimes referred to as secondary myopathies. Conditions in which this is seen include starvation, malignant disease, hypothyroid disease, infection with human immunodeficiency virus, and disorders such as pituitary and adrenal tumours associated with raised levels of circulating cortisol. The muscle wasting is generally a preferential atrophy of type 2 fibres (Fig. 15.2A), and this is also common in patients receiving prolonged or intensive treatment with anti-inflammatory steroids for asthma or inflammatory disease.

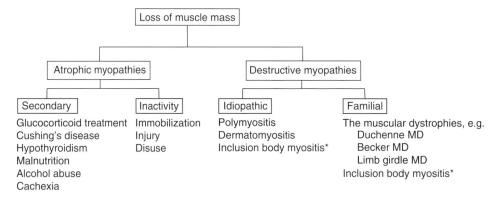

**Figure 15.1** Myopathic disorders giving rise to loss of muscle mass. *The classification of inclusion body myositis is uncertain.

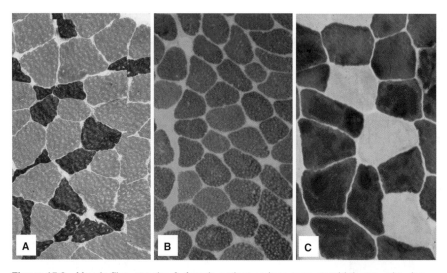

**Figure 15.2** Muscle fibre atrophy. **A**, female patient on long-term steroid therapy showing atrophic type 2 fibres. **B**, biopsy of male spinal cord injury patient, 5 years after his accident. Note the uniform fibre-type staining. **C**, normal adult male at same magnification to compare muscle fibre typing and size. All biopsies from quadriceps, stained for myosin ATPase at pH 9.4; type 2 fibres dark.

Alcoholic myopathy also leads to type 2 fibre atrophy, although this condition often includes damage to peripheral nerve fibres, which in turn leads to denervation. Occasionally excessive alcohol consumption can cause a rapid breakdown of muscle fibres and the release of soluble contents into the circulation (rhabdomyolysis).

Type 2 fibre atrophy is a common observation in muscle biopsy specimens and, as such, it is not diagnostic of any particular disease. In general, it appears that type 2 fibre atrophy is a response to metabolic stress. Type 2 fibres are used for occasional rapid movements but their other role in the body is to constitute a store of protein. As a result of malnutrition or other stresses, this protein store is called upon to provide amino acids for the maintenance of other more vital organs and tissues.

On average, type 2 fibres make up about half the bulk of most muscles and in severe cases they can shrink to almost nothing (Fig. 15.3A). In these cases, even if there were no change in the type 1 fibres, the muscle strength would be reduced to about half and the contractile properties considerably slowed. Although the atrophic fibres may become very small, they do not disappear, and if the underlying problem can be diagnosed and treated successfully the atrophy will be corrected and normal muscle fibre size and strength will be restored (Fig. 15.3B), albeit rather slowly.

Muscle atrophy may occur after disuse. Extensive and rapid loss of muscle bulk is seen if a limb is immobilized in plaster and, in contrast to what is seen with the metabolic myopathies, there is a characteristic preferential atrophy of type 1 slow twitch fibres, which occurs within a matter of weeks (see Ch. 14). Although apparent recovery of normal function may occur within 6 weeks after mobilization of the limb, residual muscle weakness may still be detectable several years after the injury unless specific strengthening exercises have been prescribed and followed (Rutherford et al 1990).

The most pronounced case of 'disuse atrophy' occurs in patients with spinal cord injury where all voluntary control of muscles below the lesion is lost (see Ch. 14). In such cases there is a conversion of the slow type 1 fibres to type 2, with all the fibres being somewhat smaller than in normal muscle (see Fig. 15.2B,C).

### Destructive myopathies

This group includes the muscular dystrophies, where the defect is genetic, and the destructive myopathies of autoimmune origin, where there is no known genetic cause but where individuals

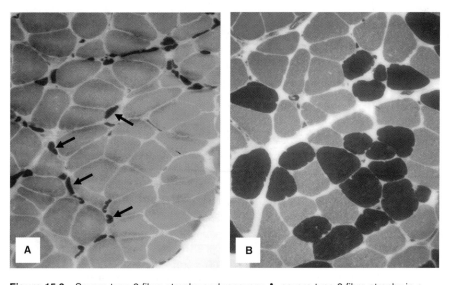

**Figure 15.3** Severe type 2 fibre atrophy and recovery. **A**, severe type 2 fibre atrophy in a patient before removal of a pituitary tumour (arrows indicate atrophic fibres). **B**, 7 months after surgery and removal of the tumour. ATPase stain, pH 9.4; type 2 fibres dark.

with certain HLA profiles (genetic backgrounds) may be more susceptible.

**The muscular dystrophies** The muscular dystrophies include all the familial (genetic) progressive disorders of muscle that result in fibre destruction and the relentless replacement of muscle cells with fat and fibrous tissue. The most serious type, although not the most common, is *Duchenne muscular dystrophy* (DMD). The defective gene is carried on the X chromosome and is recessive so that girls, who have two X chromosomes, almost always have at least one functioning gene and so act as unaffected heterozygote carriers. If, on the other hand, a boy carries the defective X chromosome, there is nothing to compensate for the deficiency and the disease is expressed. This is the sex-linked recessive mode of inheritance. Although the plasma creatine kinase (CK) levels may be as high as $10\,000\,\text{IU}\,\text{L}^{-1}$, the condition is often not noticed at first as there are few physical signs of the disease at birth. By about 4 years of age a delay in motor milestones becomes apparent and subsequently the child has increasing difficulty in standing and walking, and is usually confined to a wheelchair by the start of the second decade. As the disease progresses, weakness of respiratory muscles, often complicated by curvature of the spine (scoliosis) caused by muscle weakness, becomes a problem, and death usually occurs in the late teens or twenties.

Muscle biopsy initially shows differences in fibre size with numerous internal nuclei (Fig. 15.4A). Later, necrotic fibres become more numerous and there is some evidence of regeneration, but as the condition progresses the continuing destructive process becomes more marked until virtually no contractile material remains and the muscle consists largely of fat and fibrous tissue (Fig. 15.4B).

The defective gene in DMD was first described by Kunkel (1986) and assigned to position p21 on the short arm of the X chromosome. This very large gene (about 2000 kilobases) codes for a structural protein called dystrophin, which provides a link between the surface membrane and the underlying contractile proteins (see Fig. 1.5).

The crucial protein dystrophin is absent in the muscle fibres of patients with DMD due to deletions in the dystrophin gene giving rise to a frame shift in the DNA sequence so that either no protein or nonsense protein is produced. In the less severe Becker dystrophy, where there may be

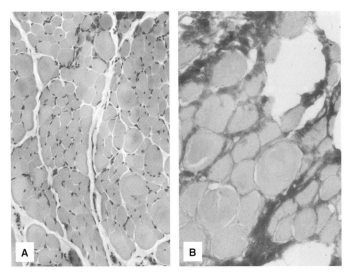

**Figure 15.4** Duchenne muscular dystrophy. **A**, early stage showing variation in fibre size and some evidence of regeneration (at the top of the section); haematoxylin and eosin stain. **B**, advanced stage showing proliferation of connective tissue and areas of fat (empty spaces). Van Gieson stain.

a deletion but no frame shift, an incomplete protein is produced with some residual function.

Although relentless and tragic, the progression of Duchenne and Becker dystrophies is relatively slow, death commonly occurring after 20–30 years. It appears that the loss of linkage between the surface membrane and the contractile apparatus leads to a cycle of continual damage and repair in which damage slowly comes to predominate.

The *mdx* mouse is an animal model of DMD, which also lacks dystrophin. However, it is interesting to note that, although young mice suffer serious dystrophic changes, their muscles regenerate. The fibres are often of varying sizes with central nuclei, and the muscles often become somewhat larger than those of normal adult mice. Regeneration appears to outstrip the degenerative process in the *mdx* mouse, whereas in human muscle it is the other way round.

With the characterization of the defective gene and gene product in Duchenne dystrophy, it is possible to provide very effective gene counselling and screening in utero where there is a family history. However, many cases occur where there is no family history as a result of spontaneous mutations in this very large gene.

Many attempts have been made to introduce the dystrophin gene into diseased muscle, most often in the form of implants of cultured healthy myoblasts, but these have met with limited success. Utrophin is a protein closely related to dystrophin; it is found widely in the body, but in mature muscle is located only at the neuromuscular and myotendinous junctions (Tinsley et al 1994). It is, however, found in regenerating muscle fibres in both dystrophic patients and normal subjects. Forced expression of full-length utrophin in the *mdx* mouse has been found to prevent the development of dystrophy (Tinsley et al 1998) and upregulation of the *utrophin* gene provides a hopeful therapeutic approach for the treatment of the human Duchene and Becker dystrophies (Wakefield et al 2000).

The dystrophin molecule is part of a complex structure linking the contractile apparatus to the surface membrane and the extracellular matrix (see Fig. 1.5). Defects in other components of this complex also give rise to a range of disorders that

are similar to the classical Duchenne and Becker dystrophies but differ in their onset, distribution and severity of the affected muscles. Defects in the sarcoglycans in the surface membrane give rise to the *limb girdle* dystrophies, while there are a number of defects of laminin that give rise to a family of disorders collectively known as the *congenital muscular dystrophies*.

In clinical practice *myotonic dystrophy* is the most likely form of dystrophy to be encountered in adults (Harper 2001). The condition is dominantly inherited with variable gene penetration, and the abnormal gene, located on chromosome 19, has been identified as coding for a protein kinase. Weakness starts peripherally and only later are proximal muscle groups involved. Myotonia of the hands is often an early symptom. The myotonia is not the result of an altered chloride conductance as is the case for congenital myotonia (see below) and it is not known how a defect in a gene coding for a protein kinase changes membrane properties leading to myotonia. Although giving its name to the disorder, myotonia itself is not the major problem. This is a multisystem disease affecting many organs, including heart and smooth muscle, and endocrine abnormalities and cataracts are common.

*Congenital myopathies* are a heterogeneous group of disorders that may manifest as a failure in the development of one particular fibre type leading to a predominance of the other (usually type 1), or the presence within the muscle fibres of abnormal structures. *Nemaline myopathy,* where tubular aggregates are seen within fibres, is a consequence of mutations in one of the thin filament proteins, while *central core disease* is due to an abnormal accumulation of the ryanodine receptor in the sarcoplasmic reticulum and may be associated with malignant hyperthermia (see below). In general, these are disorders of early childhood, which remain static or progress slowly.

**Inflammatory destructive myopathies** Polymyositis and dermatomyositis are the most commonly seen types of inflammatory myopathy. Most authorities consider them to be variants of the same disorder. Both diseases are almost certainly the result of an autoimmune process and have a female: male preponderance of about 3:1.

Plasma CK concentration is often markedly raised and a myopathic EMG is sometimes found (small irregular, asynchronous, action potentials). In severe acute polymyositis massive destruction of muscle fibres (rhabdomyolysis) may occur with the release of soluble proteins into the circulation and the threat of myoglobin precipitating and blocking the renal tubules leading to renal failure. Myoglobinuria causes the urine to become a dark brown or black colour.

The main pathological changes seen in muscle biopsies are muscle fibre necrosis with inflammatory infiltrates, mainly lymphocytes in the early stages, around blood vessels (perivascular) between and within muscle fascicles (perimysial and endomysial areas). Macrophages predominate in later stages and regenerating fibres may also be seen (Fig. 15.5). The presence of small atrophic fibres around the edge of fascicles (perifascicular atrophy) is often a feature of the condition, together with fibres with abnormal architecture. Although regenerating fibres may be present in acute cases, they are seen less often in chronic polymyositis where the response to immunosuppressive drugs has been poor. In these cases atrophy of the remaining muscle cells, particularly

of type 2 fibres, may occur due to the prolonged negative nitrogen balance produced by the anti-inflammatory steroid therapy (see Atrophic myopathies above and Fig. 15.2A).

One of the dilemmas of treating polymyositis with high-dose steroids is that the steroids themselves produce a negative nitrogen balance, causing fibre atrophy and muscle weakness which may obscure recovery from the underlying disease. In this situation careful monitoring of muscle strength and CK levels can be most helpful in managing the withdrawal of immunosuppressive medication (Edwards et al 1979).

**Mechanisms of damage in autoimmune muscle disease** It is likely that autoimmune muscle disease develops as a result of some type of environmental exposure in genetically susceptible individuals. In dermatomyositis and polymyositis there may be an attack by the immune system on muscle fibres that have undergone some change in cell surface antigens, perhaps due to a virus infection, rendering them 'foreign' to the patient's immune system. Alternatively, muscle damage may be secondary to changes in the blood vessels and capillaries that are closely associated with muscle cells, as immune reactions against the

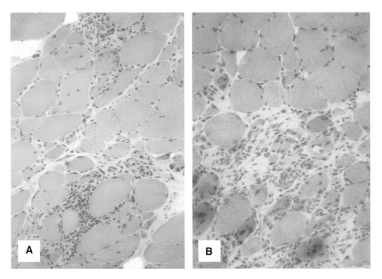

**Figure 15.5** Biopsy appearance of polymyositis. **A**, early changes with inflammatory infiltrates in perimysial and perifascicular areas. **B**, severe acute case with inflammatory cells surrounding necrotic fibres. Some evidence of regeneration, seen as dark-staining basophilic fibres. Haematoxylin and eosin stain.

endothelial cells of small blood vessels can cause inflammation and necrosis, which could account for the perivascular infiltration that is sometimes seen. As a byproduct of this inflammatory process, muscle fibres may become damaged leading to necrotic changes followed by invasion of the tissue by macrophages.

Although the autoimmune diseases are not directly 'caused' by genetic defects, several genetic risk factors have been identified as being associated with the genes that determine the HLA tissue types, and are important in signalling and recognition at the cellular level.

**Inclusion body myositis** In some cases diagnosed as polymyositis, the patient does not respond to steroids or other immunosuppressive drugs and the damage to muscle fibres becomes relentless, with a poor prognosis. Microscopic appearances in the advanced stages resemble those seen in muscular dystrophy, with replacement of contractile material with fibrous connective tissue and fat. A diagnostic finding, although not always easy to see, is the presence of rimmed vacuoles in the muscle fibres. Although previously thought of as unresponsive polymyositis, this is now generally recognized as a separate condition, designated inclusion body myositis, possibly with a familial component, hence the uncertainty about its classification in

Figure 15.1. In contrast to polymyositis where there is a female preponderance, the condition is most common in men aged over 55 years, and may be very much more common than hitherto believed. Although the condition is often sporadic with no family history, there are also familial forms. In sporadic cases the characteristic vacuoles appear to be filled with abnormal proteins, some of which are similar to those found in the cerebral plaques in Alzheimer's disease (Askanas & Engel 2001).

There is also an array of autoimmune diseases attacking connective tissue and blood vessels that affect muscle, including *multisystem vasculitis, polyarteritis nodosa, Sjögren's syndrome, rheumatoid arthritis, systemic lupus erythematosus* and *scleroderma*. For further information on these disorders, see Round & Jones (2004).

## Abnormal function with a relatively normal muscle mass

This is a heterogeneous group of relatively rare conditions where patients may present with a rapid onset of fatigue, abnormally prolonged contractions or episodes of flaccid paralysis (Fig. 15.6).

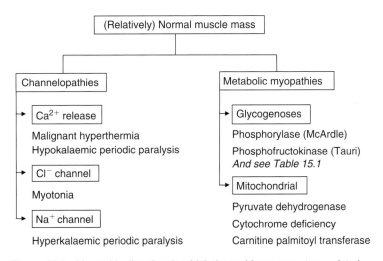

**Figure 15.6** Myopathic disorders in which the problems are not associated with a loss of muscle mass.

**Table 15.1**  Disorders of glycogen metabolism affecting skeletal muscle

| Type | Name | Deficiency | Affected organs and symptoms |
|---|---|---|---|
| 1 | Von Gierke's disease | Glucose-6-phosphatase | Hepatomegaly, growth retardation, hypoglycaemia, lactacidosis |
| 2 | Pompe's disease | Acid maltase: lysosomal enzymes $\alpha$-1,4 and $\alpha$-1,6 glucosidase | Cardiomyopathy and severe weakness with large accumulations of glycogen in the muscle |
| 3 | Cori–Forbes disease | Absence of debrancher enzyme | Abnormal glycogen accumulates in the muscle. Clinical symptoms vary from severe muscle weakness in childhood to asymptomatic adult forms |
| 5 | McArdle's disease | Myophosphorylase deficiency | Cannot metabolize muscle glycogen. Severe exercise limitation in the absence of lactic acid accumulation |
| 7 | Tauri's disease | Phosphofructokinase deficiency | Enzyme missing in muscle and red cells. Similar to McArdle's, but more severe problems with exercise. Haemolytic anaemia may occur |

## The metabolic myopathies

Muscular activity places major demands on energy metabolism, both anaerobic glycolysis and the oxidation of carbohydrates and fatty acids. All of these pathways are important in the secondary supply of energy to the muscles as the primary reserves of phosphocreatine become depleted. In general, patients with metabolic defects in these pathways are of normal, or near-normal, strength when rested but are limited to various degrees in their exercise endurance.

**The glycogenoses**  Included in Table 15.1 are diseases in which there are defects in the enzymes of the glycolytic pathway that affect skeletal muscle. In many of the glycogenoses there is an accumulation of abnormal glycogen because the enzymes breaking it down are missing, and this accumulation damages the muscle fibres.

Muscle function has been studied most extensively in patients with McArdle's disease (myophosphorylase deficiency, type 5) where the key clinical feature is of painful muscle fatigue that occurs in the absence of any accumulation of lactic acid. An ischaemic forearm exercise test is often used but this is difficult to quantify and can lead to false positives; absence of phosphorylase on histochemical staining of a muscle biopsy is diagnostic and a number of genetic defects have now been described (see below). Unless the muscle has recently gone into a contracture, it does not show any signs of damage or atrophy.

In this condition, mild exercise such as walking slowly can be sustained by the oxidation of fatty acids and blood-borne glucose. Higher-intensity exercise can be maintained for only a few seconds before the muscles begin to fatigue as the reserves of phosphocreatine are exhausted and muscle ATP levels are compromised (Cady et al 1989). If exercise continues, painful muscle contractures develop which are electrically silent (unlike muscle cramp in normal subjects, which is electrically active). These contractures, which are extremely painful, resemble rigor mortis where myosin fails to detach from the actin because of low ATP levels. The low energy status of the muscle may also allow calcium to escape from the sarcoplasmic reticulum, so compounding the problem. Contractures resolve slowly and can lead to severe muscle damage (rhabdomyolysis) when the blood supply eventually returns. Fortunately, patients are generally well aware of their limitations and take care to reduce their exercise level before any crisis can develop.

McArdle's patients differ in the severity and onset of their symptoms, which may reflect the diverse genetic defects underlying this disease (Bartram et al 1994). Some patients produce mRNA but no protein, others produce no mRNA or protein. This molecular heterogeneity implies the involvement of multiple mutations. The muscle glycogen phosphorylase gene is approximately 14 kb in size and has been assigned to the long arm of chromosome 11 (Lebo et al 1984).

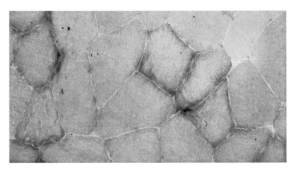

**Figure 15.7** Mitochondrial myopathy: patient with a defect in complex 1. Note the accumulations of mitochondria around the edge of the fibres. These appear red with trichrome stain and are known as 'ragged red' fibres.

**Mitochondrial myopathies** Patients with defective mitochondrial function have very limited exercise capacity, and even mild activity is associated with breathlessness and severe metabolic acidosis, such as would be seen in a normal subject exercising at high intensity or under hypoxic conditions.

Although limited in their exercise capacity, the normal glycolytic function in these patients must provide sufficient energy to prevent development of the muscle contractures that are a feature of McArdle's disease. Patients have abnormal 'ragged red' fibres in muscle biopsy sections treated with trichrome stain, the periphery of the fibres being packed with clumps of red-staining mitochondria (Fig. 15.7). Scattered necrotic fibres are also seen.

Both mitochondrial DNA and nuclear genes are necessary for the biogenesis of the respiratory chain, and pathogenic mutations of both genomes are being identified. There are now many examples where symptoms can be explained by the specific mitochondrial lesion (Larsson & Oldfors 2001; for a review see Smeitink et al 2001).

**Disorders of fat metabolism** Patients with disorders of fatty acid metabolism have symptoms of weakness, exercise intolerance, muscle stiffness and pain (sometimes accompanied by myoglobinuria). Symptoms are most evident at times when free fatty acids are the main substrates for energy metabolism, such as during prolonged submaximal exercise, particularly in the fasting state. The β-oxidation depends on the transport of free fatty acids into the mitochondria by a shuttle mechanism, which involves carnitine and the carnitine palmitoyl transferase (CPT) enzymes. Low plasma levels of carnitine and deficiencies of CPT have been described. Patients lacking carnitine are weak, whereas those lacking the CPT enzymes are of fairly normal strength at rest but after fasting or exercise may show evidence of muscle damage, with a raised CK level and occasionally myoglobinuria. Muscle biopsy may show large fat droplets in the muscle fibres.

### The ion channelopathies

Normal muscle physiology depends on the function of a variety of ion channels in the surface, T tubular and sarcoplasmic reticular membranes. A number of disorders due to mutations of cation channels have been identified and give rise to a group of diseases designated as 'ion channelopathies' (for reviews see Lehmann-Horn & Jurkat-Rott 1999, Surtees 2000).

**Malignant hyperthermia** This is a dominantly inherited disorder that can lead to emergencies during surgery where halothane is used as the general anaesthetic. The abnormal muscle responds to small doses of halothane with a prolonged contracture, which causes metabolic depletion of the muscle. The first signs of danger are muscle rigidity, a rise in the patient's temperature, and the release of potassium into the circulation, which brings with it the risk of cardiac arrest. Rapid treatment with dantrolene (which inhibits $Ca^{2+}$ release from the sarcoplasmic reticulum) and cooling are effective. If the patient survives the acute emergency, the severely damaged skeletal muscles release soluble proteins and there is then the risk of acute renal failure as the myoglobin blocks the renal tubules.

Malignant hyperthermia is now known to be associated with a group of over 20 mutations in the ryanodine receptor (see Ch. 3) (Jurkat-Rott et al 2000). The anaesthetic halothane acts in a very similar way to caffeine in that high doses will open ryanodine channels leading to the release of calcium from the sarcoplasmic reticulum. In most people the anaesthetic dose is never high enough to cause a problem, but individuals with malignant hyperthermia are particularly sensitive and in these cases the effects are dramatic and life threatening. In some families malignant hyperthermia is associated with central core disease (see above) and in vitro tests are recommended for these patients and relatives if surgery is contemplated (Curran et al 1999). A similar defect became apparent in Danish pigs that had been selectively bred to produce very lean meat, although the contractures in pig muscle seem to be precipitated by a wide range of stresses. Much of the research into this disease was stimulated by the commercial problems of the Danish pig industry.

**Myotonia congenita** In this condition there is a defect in the gene encoding the major skeletal muscle chloride channel (Lehmann-Horn & Jurkat-Rott 1999). Chloride channels have the effect of reducing membrane excitability, which is important in skeletal muscle where a voltage gradient can build up between the surface membrane and the depths of T tubules. Consequently a low chloride conductance gives rise to excitable membranes that fire repetitively having been stimulated once, and this leads to a prolonged contraction. The EMG signal from a normal muscle stops abruptly when the voluntary action ceases, but with myotonia the electrical discharge continues, fading away after about 30 s. The physiological basis of this condition was first worked out in a strain of goats that had the spectacular trick of falling over with rigid muscles whenever they were startled (McComas 1996). It is suggested that this strain was selectively bred by poor American farmers because the goats were unable, or unwilling, to jump over fences.

**Periodic paralyses** There are two forms of periodic paralysis, the hyperkalaemic and hypokalaemic varieties in which attacks of flaccid paralysis are provoked, respectively, by high or low plasma potassium concentrations (Jurkat-Rott et al 2000). With hyperkalaemic periodic paralysis the defect is of sodium channels that go into an abnormal refractory state when the surface membrane is depolarized. The hypokalaemic variety is a defect in the calcium channels forming the dihydropyridine (DHP) receptors that become inactive in the presence of low external potassium concentrations and thus ineffective in linking electrical activity with release of calcium from the sarcoplasmic reticulum. Restoring normal blood potassium levels reverses the paralysis in both cases but it is important to diagnose the problem correctly as giving potassium to a patient with the hyperkalaemic variety can only exacerbate the condition. High blood potassium levels can be reduced by giving glucose and/or insulin.

## FATIGUE AND DISEASE STATES

Fatigue is one of the most common presenting symptoms and it is natural, when looking for explanations, to think in terms of glycolytic or mitochondrial abnormalities as described above. In reality, problems such as McArdle's disease are rare and mitochondrial defects even rarer. The majority of patients in whom the muscle bulk is relatively well preserved suffer from fatigue of a central nature rather than from peripheral metabolic problems in their muscles, vascular disease being a possible exception although here the problem is more one of ischaemic muscle pain.

**Chronic heart failure** Fatigue is an almost universal experience in these patients. As the basic problem is a reduced central pump activity, it would seem self-evident that a reduced blood supply to working muscles is the explanation of these feelings. However, there are many cases where limb blood flow is normal but exercise tolerance is reduced (Wilson et al 1993) and there is little evidence of peripheral muscle abnormality (Buller et al 1991). A survey of 52 ambulatory patients found no relationship between the feelings of fatigue and the severity of the cardiac impairment (Wilson et al 1995), indicating a complex situation.

**Rheumatoid and endocrine diseases** In patients with *rheumatoid arthritis, osteoarthritis* and *fibromyalgia*, symptoms of fatigue are common but poorly related to the severity of the physical disease, with pain and sleep disturbance being the factors that correlate most closely with fatigue (Wolfe et al 1996).

Patients with both *hypothyroid* and *hyperthyroid* disease complain of fatigue. Hypothyroid patients are especially interesting in the present context because, despite fatigue being one of their main complaints, their muscles, when tested objectively, prove to be more resistant to fatigue than normal (Wiles et al 1979).

**Basal ganglia** Feelings of fatigue are also a common complaint of patients suffering from Parkinson's disease, and there is growing interest in the possibility that abnormalities in the basal ganglia may increase perception of the effort required to initiate movement (Chaudhuri & Behan 2000).

**Chronic fatigue states** Chronic fatigue syndrome, or myalgic encephalitis (ME), is a complex condition, or group of conditions, with one common factor: an abnormally low tolerance of whole body exercise (White 1990). Whilst there may well be alterations in cardiovascular and muscle function, evidence indicates that such changes are probably secondary to decreased activity and are not its underlying cause. Consequently recent attention has turned to possible central nervous system changes, such as the role of the brainstem in regulating levels of attention and arousal and increasing sensitivity to the sensations of exercise. An increase in the sensitivity of brain 5-hydroxytryptamine receptor function is associated with chronic feelings of fatigue and exercise intolerance (Bakheit et al 1992, Cleare et al 1995), implicating pathways in the hypothalamus as possible sites where abnormalities may occur.

# NEUROPATHIES

Voluntary skeletal muscle cannot function without an intact nerve supply, and in terms of clinical importance the most common muscle disorders are those that are secondary to a defect in the nervous system. Three groups of neurogenic muscle disease can be identified according to the site of the lesion: (1) lesions in the anterior horn cells, as in the *spinal muscular atrophies* and *motor neuron disease*; (2) axonal lesions, as in *peripheral neuropathies*; and (3) lesions at the motor endplate as in *myasthenia gravis* (Fig. 15.8).

**Muscle biopsy appearance** The microscopic appearance of a muscle biopsy specimen can give important information about the differential diagnosis and the progress of the neuropathic disorders. The muscle fibres constituting a single motor unit are widely distributed throughout the muscle, so that loss of a motoneuron or damage to an axon or its peripheral branches will result in the appearance of scattered atrophic muscle fibres. These fibres atrophy because they no longer receive a nervous input and do not contract. These are often referred to as *small angular fibres* (Fig. 15.9).

In relatively mild cases the angular fibres may be occasional and scattered around the section; however, if the condition progresses, more fibres are affected, some of which are adjacent, constituting 'small group atrophy'. As adjacent fibres will usually be innervated by different motoneurons, the atrophic group will consist of both fibre types, which differs from the selective atrophy of individual fibre types seen with metabolic conditions or immobilization, as described above. A further development is of 'large group atrophy' where, as the name implies, large groups of fibres are

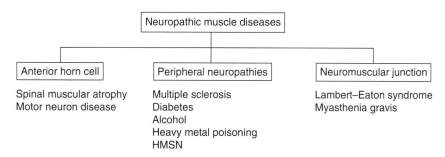

**Figure 15.8** Muscle disorders with a neuropathic origin. HMSN, hereditary motor and sensory neuropathy.

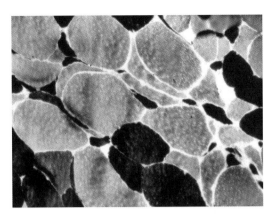

**Figure 15.9** Small angular fibres as a result of denervation. Note that both types of fibre are affected. Quadriceps biopsy, ATPase stain; type 2 fibres appear dark.

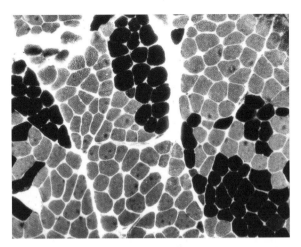

**Figure 15.11** Fibre-type grouping in a quadriceps muscle from a woman aged 42 years who suffered peripheral nerve damage from heavy metal poisoning. Groups of one fibre type indicate reinnervation. ATPase stain, pH 9.4; type 2 fibres stain dark.

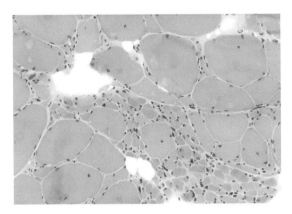

**Figure 15.10** Large group atrophy in a patient with motor neuron disease. Quadriceps biopsy stained with haematoxylin and eosin. Note some relatively normal fibres, although often with central nuclei, and a swathe of small atrophic fibres stretching from the bottom right-hand side to the centre of the section.

affected (Fig. 15.10). In some cases, most notably in motor neuron disease, this may affect one whole fascicle, which raises a host of questions about the aetiology of the condition. If the condition arises as a result of a defect in the motoneuron, the consequences would be expected to be seen diffusely throughout the muscle, covering the territory of that motor unit and not restricted by the boundary of a fascicle.

Small and large group atrophy, where both fast and slow fibre types are affected, probably indicates a fairly acute state of the disease process: if the nerve damage is chronic, evidence of reinnervation appears, leading to fibre-type grouping.

Mature muscle fibres are innervated by a single axon, but if the innervation is lost, nearby undamaged axons sprout and branch to form new contacts with the denervated muscle fibre. Initially there may be multiple innervation, which undergoes the same process of selection and elimination as seen in fetal muscle (see Ch. 14). If the process of denervation and reinnervation occurs and reoccurs over a long period, one healthy axon comes to dominate an area of muscle and all the fibres assume the contractile and histochemical characteristics set by its motoneuron. This change is seen in muscle biopsy sections as groups of muscle fibres that are all of the same fibre type (Fig. 15.11), in contrast to the random arrangement of fibre types seen in healthy muscle.

This arrangement of fibres gives rise to a characteristic pattern of electrical activity. Normally adjacent fibres are from different motor units and they fire independently and asynchronously. When recorded with surface or needle electrodes this gives a random moderate level of electrical activity. However, when a group of adjacent fibres are all innervated by the same motoneuron they all fire together and this is seen as a series of 'giant' action potentials.

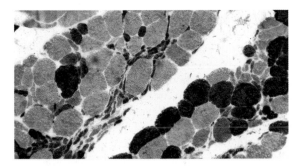

**Figure 15.12** Motor neuron disease showing small group atrophy of both fibre types. ATPase stain, pH 9.4; type 2 fibres stain dark.

## Disorders of the spinal motoneurons

**Motor neuron disease** *(amyotrophic lateral sclerosis)* This is a disease of middle to old age, responsible for about 1 in 1000 adult deaths, with a male predominance, and is characterized by widespread degeneration of motoneurons. The mechanism of this degeneration is not known but recent studies suggest that it may be a form of abnormal programmed cell death (apoptosis) (Martin et al 2000).

The first clinical signs often appear in the small muscles of the hand, and in most cases the disease progresses rapidly to a global weakness. Wasting and fasciculation of the limb muscles with spasticity and brisk reflexes is characteristic of the disease, and the duration from clinical onset to death is rapid, rarely exceeding 3 years. The majority of cases are sporadic, although familial forms exist. In these latter cases a number of genes have been implicated, including that for *superoxide dismutase 1*, which is suggestive of a role for free radicals in the disease process (Orrell 2000). Muscle biopsy shows small group atrophy of both fibre types (Fig. 15.12).

The disease is unusually common in three distinct geographical locations: the Pacific island of Guam, the Kiwi peninsula of Japan and New Guinea. This has led to speculation about the involvement of specific environmental factors including the toxic effects of manganese, aluminium and lead. The consumption of certain plant neurotoxins has also been discussed. Flying foxes forage on neurotoxic cycad seeds and these, in turn, are eaten by the Chammono people of Guam,

possibly leading to cumulative doses of plant toxins (Cox & Sacks 2002). The introduction of American dietary habits to these areas has been associated with a reduction in the incidence of the disease. Neurotropic viruses have been sought, unsuccessfully, and an autoimmune component has been suggested. It is also possible that some toxin is transported up the axon from the periphery to the cell body.

Another explanation for motor neuron disease is that motoneurons require continual stimulation by trophic substances secreted by the muscle and carried along the motor axon back to the cell body. It is suggested that failure of this process leads to death of the motoneuron. This is seen as analogous to the extensive loss of motoneurons that occurs in the fetus when superfluous axonal terminations are eliminated from the maturing muscle fibres, but how or why this should suddenly happen in an apparently healthy middle-aged man is not known.

**Poliomyelitis** This is an acute acquired disease of the motoneurons that has been almost eradicated from the world as a result of mass immunization. However, in the 1950s there were a series of major epidemics in Europe and America, predominantly affecting young children, hence its alternative name of *infantile paralysis*. The disease is a viral infection in which there is destruction of motoneurons leading to profound muscle weakness. Other viruses, for example the Coxsackie viruses, can also invade anterior horn cells, giving rise to muscle weakness and paralysis. This disease is similar to the condition of *distemper* in dogs.

**Spinal muscular atrophy (hereditary motor neuropathy)** The spinal muscular atrophies are a group of inherited clinically and genetically heterogeneous disorders characterized by degeneration of anterior horn cells, the motor nuclei in the brainstem and, rarely, the upper motoneuron pathways. In recent years the view that several separate genes are involved has been strengthened by large family studies. Although the precise nature of the defect is not clear, it is evident that in these diseases there is some genetic defect causing the motoneuron to fail in its task of innervating and supporting muscle fibres. Autosomal recessive

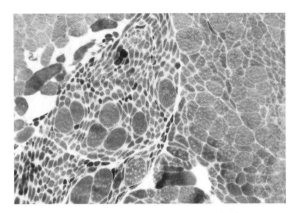

**Figure 15.13** Acute infantile spinal muscular atrophy. Note the severe fibre atrophy with scattered hypertrophied fibres. ATPase stain, pH 9.4; type 2 fibres stain dark.

spinal muscular atrophy (SMA) is the commonest form; the condition is one of the many causes of a floppy infant and is due to a mutation in a gene encoding a protein that appears to play a critical role in RNA metabolism and also interacts with mediators of programmed cell death (apoptosis). The chromosomal locations of some of the rarer mutations, including dominant forms of SMA, have also been reported (Talbot & Davies 2001). Muscle biopsy in SMA shows severe group atrophy, and sometimes whole fascicles are composed of tiny rounded fibres, which the ATPase reaction shows to be of both fibre types. In other fascicles, groups of normal fibres often with a few grossly hypertrophied fibres (Fig. 15.13) are seen, the latter often being of one fibre type.

## Peripheral neuropathies

Peripheral neuropathies are diseases characterized by structural damage and functional loss in the peripheral nervous system. The damage may be due to physical trauma, inflammatory disease, metabolic factors (e.g. diabetes where glycosylation of myelin proteins leads to problems) or toxic substances (e.g. alcohol abuse or poisoning with heavy metals), or may be genetic in origin. The incidence of peripheral nerve disease is probably higher than that of primary muscle disease. Impairment of both sensory and motor

function can occur, although one or the other may dominate the clinical picture. The peripheral nerve pathology may be the result of demyelination or of axonal degeneration. In the demyelinating neuropathies there is usually a marked slowing of nerve conduction, whereas in axonal degeneration nerve conduction is normal, provided some conducting axons remain. Distal segments of the limbs are usually most severely affected. There is a vast pathological spectrum and two of the commonest are discussed below.

**Multiple sclerosis** This is a demyelinating disease of unknown cause, although viral and autoimmune mechanisms have been suggested. There is a degree of peripheral muscle weakness as patients find it difficult to activate their muscles fully. The reduced conduction velocity along the peripheral nerve means that the muscles are not activated by the higher frequencies during voluntary efforts. This problem probably does not account for the symptoms of fatigue and loss of coordination, which fluctuate in severity and are almost certainly associated with the effects of the disease on the central nervous system.

**Hereditary motor and sensory neuropathy (HMSN)** In this syndrome, a convenient classification is by the velocity of peripheral nerve conduction. In *HMSN type I*, a low conduction velocity is associated with segmental demyelination and hypertrophy of the nerves that can be palpated (onion bulb). This is a result of Schwann cell proliferation accompanying remyelination. Onset is usually within the first 10 years of life. Normal conduction velocity is seen in *HMSN type II*, in which there is axonal degeneration with only minimal demyelination. Onset after 40 years of age is quite common in this form, which may be associated with upper motoneuron lesions or the clinical picture of Friedreich's ataxia. Patients may develop *inverted champagne bottle* or *stork* legs (fat thighs, thin calves).

Muscle biopsy of the lower limb muscles shows signs of mild denervation atrophy in the early stages of the disease, with reinnervation occurring later. There may be compensatory hypertrophy of the surviving fibres.

**Acute acquired peripheral neuropathy** The *Guillain–Barré syndrome* is probably the commonest acquired peripheral neuropathy associated with muscle weakness. Typically, an acute febrile illness is followed by generalized weakness of both proximal and distal muscles, which may become so severe that respiratory muscles are affected and assisted ventilation is required. The condition is due to inflammatory demyelination of segments of the motor nerves, which appears on nerve conduction examination as a slowing or complete block of nerve conduction. In most cases remyelination occurs and the patient recovers, although some may be left with residual weakness. Relapses occasionally occur and in some rare cases the problems persist and may spread to involve the autonomic nervous system.

## Diseases of the neuromuscular junction

**Myasthenia gravis** Myasthenia gravis is a disease characterized by the development of abnormal fatigue and fluctuating weakness. After a period of rest the muscles are of relatively normal strength, but weakness develops after only brief activity. All muscles of the body may be involved but weakness of the facial muscles with drooping eyelids often gives the patient a characteristic appearance. There may be difficulty with speech and swallowing, and involvement of the respiratory muscles can prove fatal. Onset can be at any age and may be rapid or insidious, but is most common in the third and fourth decades of life. Twice as many females as males are affected. Although muscle bulk is initially normal, inactivity, or lack of ability to maintain high-force contractions, eventually leads to muscle wasting.

A diagnostic feature is a dramatic improvement in muscle strength after injection of the cholinesterase inhibitor *edrophonium chloride* (the 'Tensilon test'). Another test is to stimulate the muscle at a low frequency, such as 3 Hz, and measure the size of the evoked action potential. Normally the amplitude of the action potential is well maintained but in myasthenia a progressive decrease in size is seen.

There is normally a reduction in the quantal release of acetylcholine during sustained muscle contraction, and in normal muscle this is of little consequence as there is a large margin of safety. In myasthenia, however, the natural transmitter is competing with antibodies to the acetylcholine receptor, and so the fall in transmitter release which occurs during sustained activity causes the number of miniature end-plate potentials (MEPPs) generated to fall below the threshold required to initiate a propagated action potential in the muscle fibre. The cholinesterase inhibitors work by maintaining a high concentration of acetylcholine in the synaptic cleft, effectively competing with the antibody.

The disease is an autoimmune process in which circulating antibodies can be demonstrated. These antibodies have two actions: the first is a curare-like inhibition of the neuromuscular junction, as described above; the second is to affect the turnover of acetylcholine receptors, giving rise to abnormally flattened postsynaptic membranes with few junctional folds, which further reduces the function of the neuromuscular junction. The initial trigger for the autoimmune response is not known. Thymectomy is often helpful if carried out early in the disease, but later, when antibody production is at extrathymic sites, thymectomy is ineffective. Plasmapheresis to remove circulating antibody can give temporary relief but long-term management of the autoimmune condition is usually with steroids or other immunosuppressive drugs. Anticholinesterase inhibitors can relieve the myasthenic symptoms and it is important to find a preparation that will give relief during periods of activity but does not have toxic side-effects.

**The Lambert–Eaton myasthenic syndrome** The Lambert–Eaton syndrome is less common than myasthenia gravis; it is characterized by weakness and fatigue, and may also be associated with abnormalities of the autonomic nervous system. The disease is more common in men and can present at any age, but in the older age groups it is often associated with malignancy. EMG findings indicate the presence of a block in neurotransmission, which, unlike myasthenia, improves with continuing activity. The defect appears to be due to an autoimmune process and antibodies have been detected against the voltage-activated calcium channels in the motor nerve that allow calcium entry and cause synaptic vesicles to fuse with the presynaptic membrane (McComas 1996). The condition responds well to corticosteroid therapy. The syndrome is often associated with

carcinoma, especially of the lung, and it is suggested that antibodies against calcium channels in the tumour also attack channels in the motor nerve terminal. Although muscle bulk is initially normal, inactivity eventually leads to muscle wasting.

## REFERENCES AND FURTHER READING

Anderson J R 1985 Atlas of skeletal muscle pathology. Lancaster: MTP Press

Askanas V, Engel W K 2001 Inclusion-body myositis: newest concepts of pathogenesis and relation to ageing in Alzheimer disease. Journal of Neuropathology and Experimental Neurology 60: 1–14

Bakheit A M O, Behan P O, Dinan T G, Gray C E, O'Keane V 1992 Possible upregulation of hypothalamic 5-hydroxytryptamine receptors in patients with postviral fatigue syndrome. British Medical Journal 403: 1010–1012

Bartram C, Edwards R H T, Clague J, Benyon R J 1994 McArdle's disease: a rare frameshift mutation of the muscle glycogen phosphorylase gene. Biochimica et Biophysica Acta 1226: 341–343

Buller N P, Jones D A, Poole-Wilson P A 1991 Direct measurement of skeletal muscle fatigue in patients with chronic heart failure. British Heart Journal 65: 20–34.

Cady E B, Jones D A, Lynn J, Newham D J 1989 Changes in force and intracellular metabolites during fatigue of human skeletal muscle. Journal of Physiology 418: 311–325

Chaudhuri A, Behan P O 2000 Fatigue and the basal ganglia. Journal of the Neurological Sciences 179: 34–42

Cleare A J, Bearn J, Allain T et al 1995 Contrasting neuroendocrine responses in depression and chronic fatigue syndrome. Journal of Affective Disorders 34: 283–289

Cox P A, Sacks O W 2002 Cycad neurotoxins, consumption of flying foxes and ALS–PDCC disease in Guam. Neurology 58: 956–959

Curran J L, Hall W J, Halsall P J et al 1999 Segregation of malignant hyperthermia, central core disease and chromosome 19 markers. British Journal of Anaesthesia 83: 217–222

Edwards R H T, Wiles C M, Round J M, Jackson M J, Young A 1979 Muscle breakdown and repair in polymyositis: a case study. Muscle and Nerve 2: 223–228

Harper P S 2001 Myotonic dystrophy. In: Karpati G, Hilton-Jones D, Griggs R C (eds) Disorders of voluntary muscle. 7th edn, pp 541–559. Cambridge: Cambridge University Press

Jurkat-Rott K, McCarthy T, Lehmann-Horn F 2000 Genetics and pathogenesis of malignant hyperthermia. Muscle and Nerve 23: 4–17

Karpati G, Hilton-Jones D, Griggs R C (eds) 2001 Disorders of voluntary muscle. 7th edn. Cambridge: Cambridge University Press

Kunkel L M 1986 Analysis of deletions in DNA from patients with Becker and Duchenne muscular dystrophy. Nature 322: 73–77

Larsson N G, Oldfors A 2001 Mitochondrial myopathies. Acta Physiologica Scandinavica 171: 385–393

Lebo R V, Gorin F, Fletterick R J et al 1984 High resolution chromosome sorting and DNA spot block analysis assign McArdle's syndrome to chromosome 11. Science 225: 57–59

Lehmann-Horn F, Jurkat-Rott K 1999 Voltage-gated ion channels and hereditary disease. Physiological Reviews 79: 1317–1372

Martin L J, Price A C, Kaiser A, Shaikh A Y, Liu Z 2000 Mechanisms for neuronal degeneration in amyotrophic lateral sclerosis and in models of motor neurone disease (review). International Journal of Molecular Medicine 5: 3–13

McComas A J 1996 Skeletal muscle: form and function. Champaign, Illinois: Human Kinetics

Orrell R W 2000 Amyotrophic lateral sclerosis: copper/zinc superoxide dismatase (SOD 1) gene mutations. Neuromuscular Disorders 10: 63–68

Round J M, Jones D A 2004 Skeletal muscle physiology in health and disease. In: Woo P, Eisenberg D A (eds) Textbook of rheumatology, Ch 3.7. Oxford: Oxford University Press

Rutherford O M, Jones D A, Round J M 1990 Long lasting unilateral muscle wasting and weakness following injury and immobilisation. Scandinavian Journal of Rehabilitation Medicine 22: 33–37

Smeitink J, van den Heuvel L, DiMauro S 2001 The genetics and pathology of oxidative phosphorylation. National Review of Genetics 2: 342–352

Surtees R 2000 Inherited ion channel disorders. European Journal of Paediatrics 159: 199–203

Talbot K, Davies K E 2001 Spinal muscular atrophy. Seminars in Neurology 21: 189–197

Tinsley J, Blake D A, Zuellig R A, Davies K 1994 Increasing complexity of the dystrophin-associated protein complex. Proceedings of the National Academy of Sciences 591: 8307–8313

Tinsley J, Deconinck N, Fisher R et al 1998 Expression of full length eutrophin prevents muscular dystrophy in *mdx* mice. Nature Medicine 4: 1441–1444

Wakefield P M, Tinsley J M, Wood M J, Gilbert R, Karpati G, Davies K E 2000 Prevention of the dystrophin phenotype in dystrophin/utrophin deficient muscle following adenovirus mediated transfer of a utrophin minigene. Gene Therapy 7: 201–204

White P 1990 Fatigue and chronic fatigue syndromes. In: Bass C. (ed.) Somatization: physical symptoms and psychological illness, pp 104–140. Oxford: Blackwell Scientific

Wiles C M, Young A, Jones D A, Edwards R H T 1979 Relaxation rate of constituent muscle-fibre types in human quadriceps. Clinical Science 56: 47–52

Wilson J R, Mancini D M, Dunkman W B 1993 Exertional fatigue due to skeletal muscle dysfunction in patients with heart disease. Circulation 87: 470–475

Wilson J R, Rayos G, Gothard P, Bak K 1995 Dissociation between exertional symptoms and circulatory function in patients with heart failure. Circulation 92: 47–53

Wolfe F, Hawley D J, Wilson K 1996 The prevalence and meaning of fatigue in rheumatic disease. Journal of Rheumatology 23: 1407–1471

# Index